Chapter 1
Number Concepts

Houghton Mifflin Harcourt

Made in the United States
Text printed on 100% recycled paper

Houghton
Mifflin
Harcourt

Printed in the U.S.A.

ISBN 978-0-544-29547-6

19 0928 20

4500800118 B C D E F G

Dear Students and Families,

Welcome to **Go Math!**, Grade 2! In this exciting mathematics program, there are hands-on activities to do and real-world problems to solve. Best of all, you will write your ideas and answers right in your book. In **Go Math!**, writing and drawing on the pages helps you think deeply about what you are learning, and you will really understand math!

By the way, all of the pages in your **Go Math!** book are made using recycled paper. We wanted you to know that you can Go Green with **Go Math!**

Sincerely,

The Authors

Made in the United States
Text printed on 100% recycled paper

GO MATH!

Authors

Juli K. Dixon, Ph.D.
Professor, Mathematics Education
University of Central Florida
Orlando, Florida

Edward B. Burger, Ph.D.
President, Southwestern University
Georgetown, Texas

Steven J. Leinwand
Principal Research Analyst
American Institutes for Research (AIR)
Washington, D.C.

Matthew R. Larson, Ph.D.
K-12 Curriculum Specialist for Mathematics
Lincoln Public Schools
Lincoln, Nebraska

Martha E. Sandoval-Martinez
Math Instructor
El Camino College
Torrance, California

Contributor

Rena Petrello
Professor, Mathematics
Moorpark College
Moorpark, California

English Language Learners Consultant

Elizabeth Jiménez
CEO, GEMAS Consulting
Professional Expert on English Learner Education
Bilingual Education and Dual Language
Pomona, California

Number Sense and Place Value

Critical Area

Critical Area Extending understanding of base-ten notation

Number Concepts 9

COMMON CORE STATE STANDARDS

2.OA Operations and Algebraic Thinking
Cluster C Work with equal groups of objects to gain foundations for multiplication.
2.OA.C.3

2.NBT Number and Operations in Base Ten
Cluster A Understand place value.
2.NBT.A.2, 2.NBT.A.3

GO DIGITAL

Go online! Your math lessons are interactive. Use *i*Tools, Animated Math Models, the Multimedia *e*Glossary, and more.

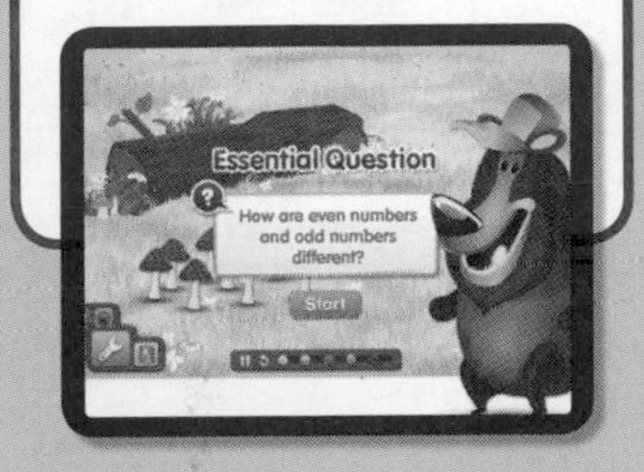

Chapter 1 Overview

In this chapter, you will explore and discover answers to the following **Essential Questions**:

- How do you use place value to find the values of numbers and describe numbers in different ways?
- How do you know the value of a digit?
- What are some different ways to show a number?
- How do you count by 1s, 5s, 10s, and 100s?

Personal Math Trainer
Online Assessment and Intervention

FOR MORE PRACTICE GO TO THE
Personal Math Trainer
Practice and Homework
Lesson Check and Spiral Review in every lesson

Whales

by John Hudson

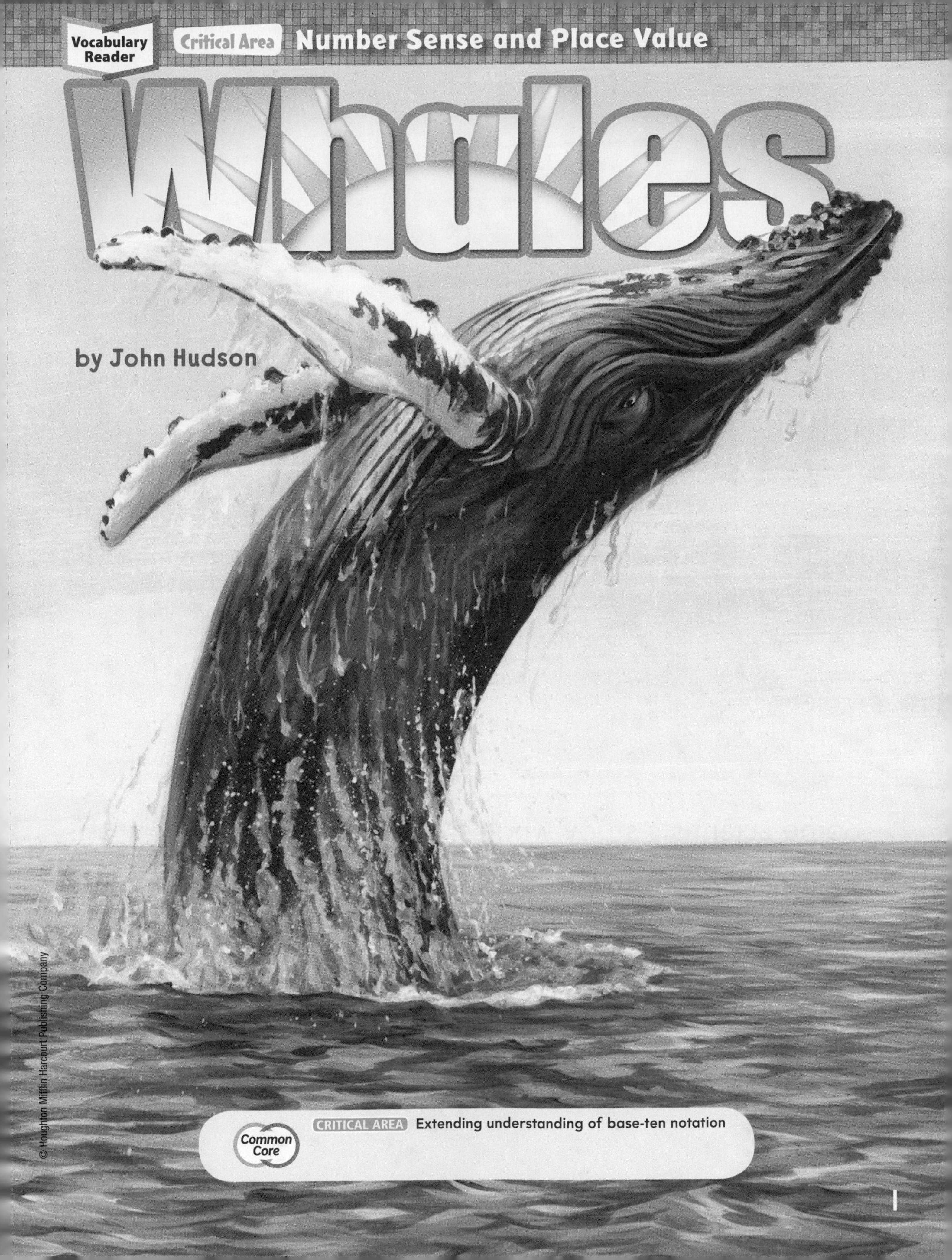

CRITICAL AREA Extending understanding of base-ten notation

Common Core

Some scientists study whales. Different kinds of whales swim along the west coast of the United States of America.
A scientist sees 8 blue whales.
Blue whales are the largest animals on Earth.

Social Studies

Where is the United States of America on the map?

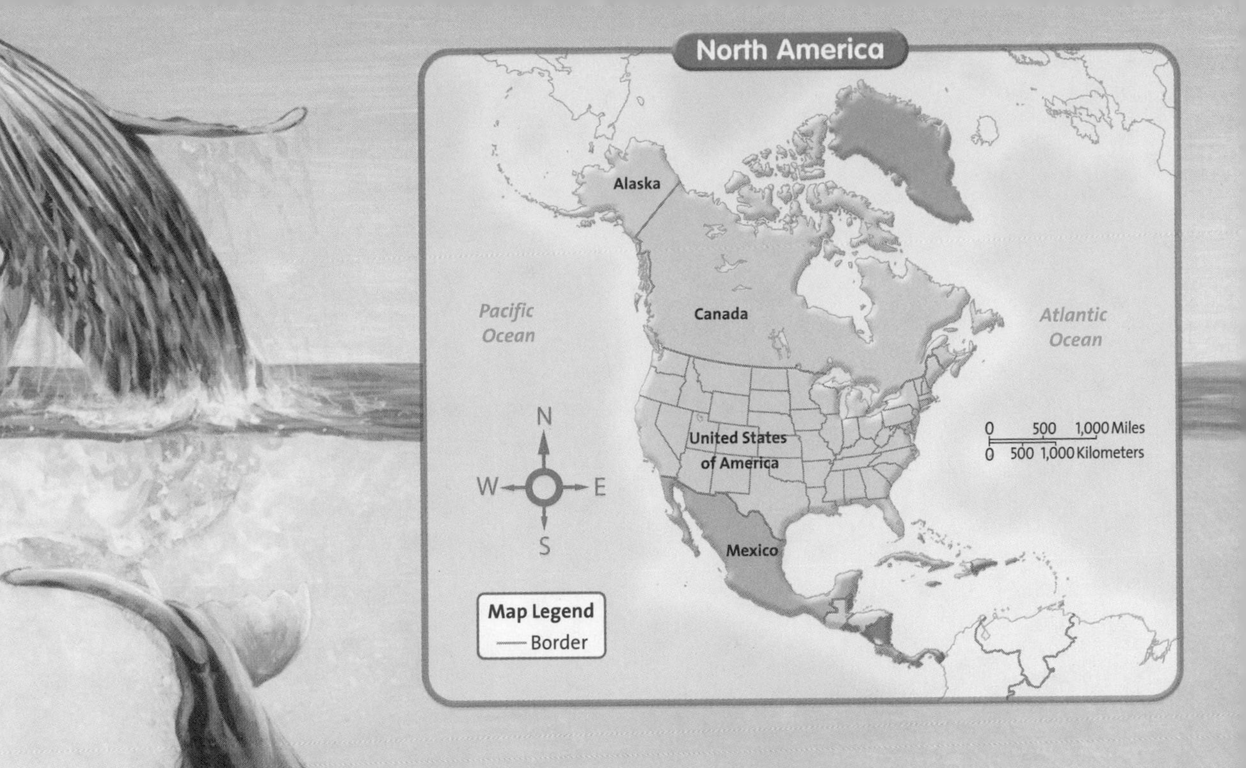

The scientist also sees 13 humpback whales.
Humpback whales sing underwater.
Did the scientist see more humpback whales or more blue whales? more ________________ whales

Social Studies

Where is the Pacific Ocean on the map?

Whales also swim along the east coast of Canada and the United States of America. Pilot whales swim behind a leader, or a *pilot*. A scientist sees a group of 29 pilot whales.

Social Studies

Where is Canada on the map?

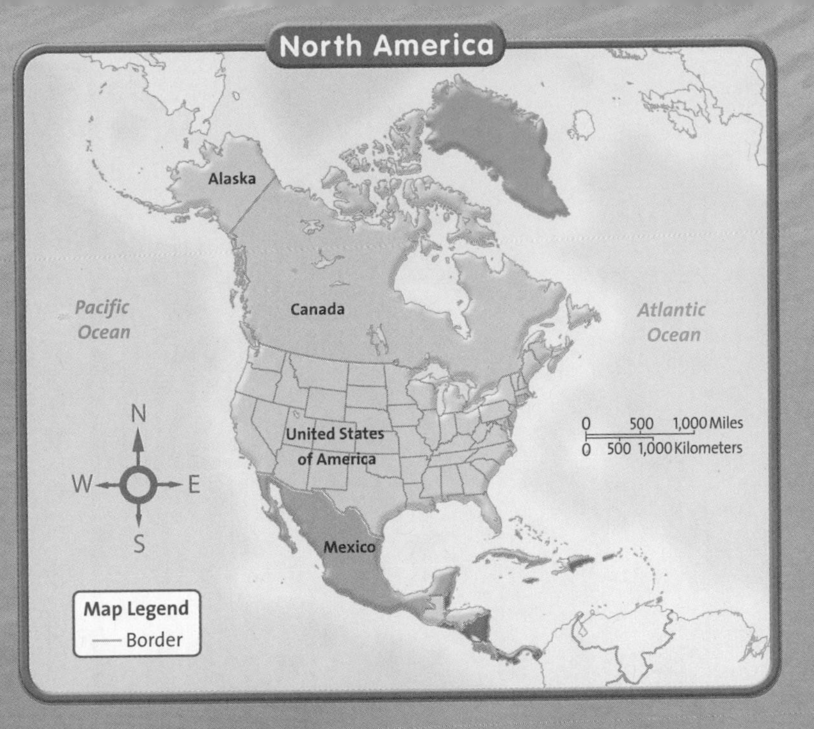

Fin whales are fast swimmers. They are the second-largest whales in the world. A scientist sees a group of 27 fin whales. How many tens are in the number 27?

_____ tens

Social Studies

Where is the Atlantic Ocean on the map?

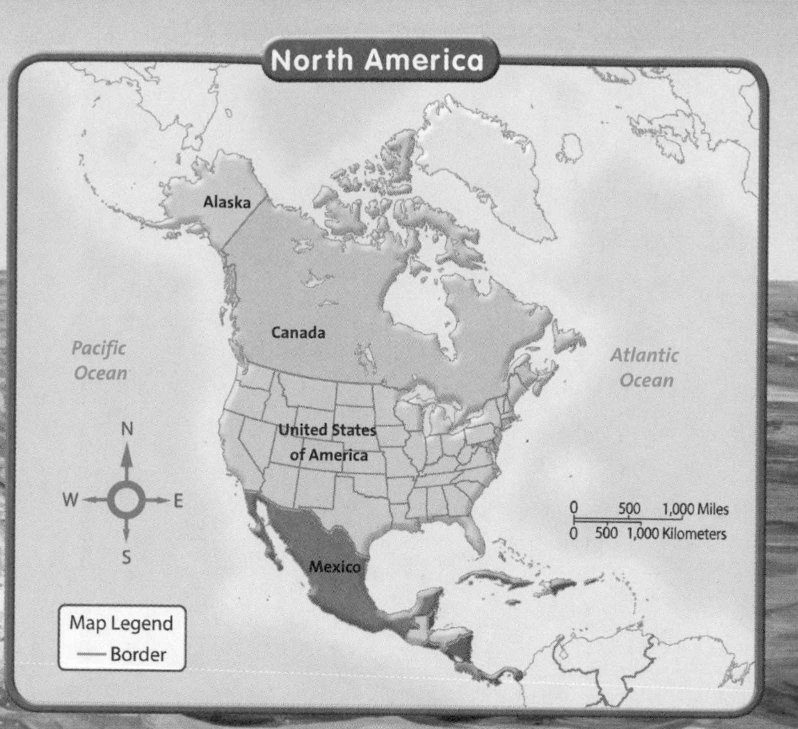

Humpback whales swim to the warm water near Mexico for the winter. Humpback whales may have as many as 35 throat grooves.

In the number 35, the _____ is in the ones place and the _____ is in the tens place.

Social Studies

Where is Mexico on the map?

Name ______________________________

Write About the Story

Look at the pictures. Draw and write your own story. Compare two numbers in your story.

Vocabulary Review

more	fewer
tens	greater than
ones	less than

____

____

____

____

____

____

The Size of Numbers

The table shows how many young whales were seen by scientists.

Young Whales Seen	
Whale	Number of Whales
Humpback	34
Blue	13
Fin	27
Pilot	43

1. Which number of whales has a 4 in the tens place?

____

2. How many tens and ones describe the number of young blue whales seen?

_____ ten _____ ones

3. Compare the number of young humpback whales and the number of young pilot whales seen. Write > or <.

34 ◯ 43

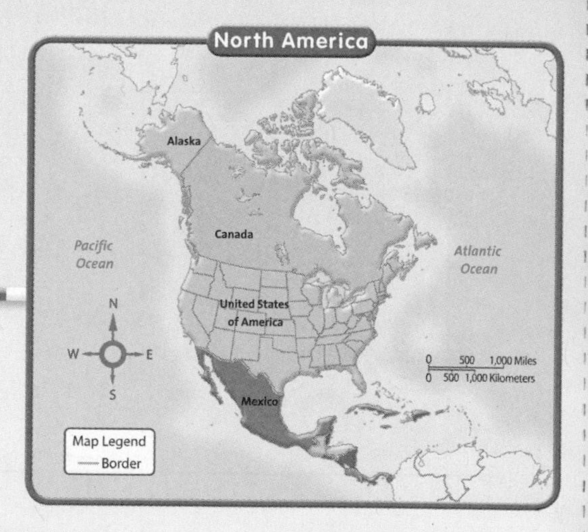

4. Compare the number of young fin whales and the number of young blue whales seen. Write > or <.

27 ◯ 13

Write a story about a scientist watching sea animals. Use some 2-digit numbers in your story.

Chapter 1

Number Concepts

At a farmers' market, many different fruits and vegetables are sold.

If there are 2 groups of 10 watermelons on a table, how many watermelons are there?

Name ______________________

Model Numbers to 20

Write the number that tells how many. (K.NBT.A.1)

1. 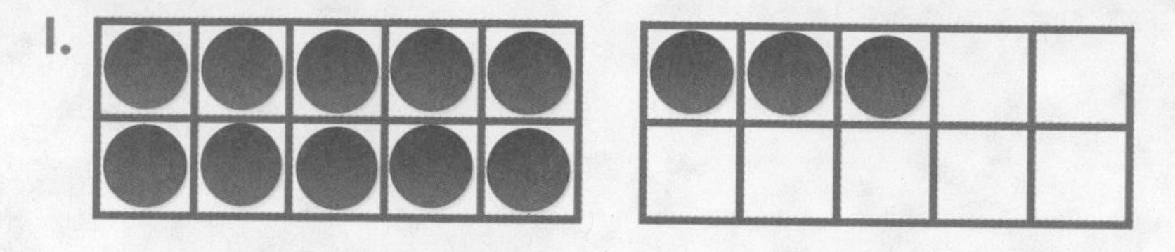

_____

2. 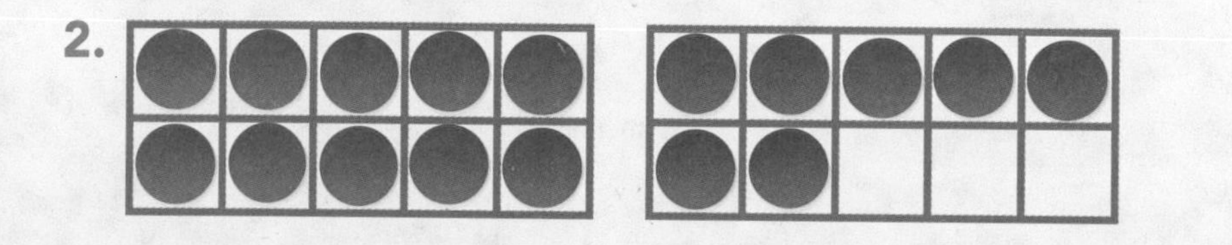

_____

Use a Hundred Chart to Count

Use the hundred chart. (1.NBT.A.1)

1	2	3	4	5	6	7	8	9	10
11	12	13	14	15	16	17	18	19	20
21	22	23	24	25	26	27	28	29	30
31	32	33	34	35	36	37	38	39	40
41	42	43	44	45	46	47	48	49	50
51	52	53	54	55	56	57	58	59	60
61	62	63	64	65	66	67	68	69	70
71	72	73	74	75	76	77	78	79	80
81	82	83	84	85	86	87	88	89	90
91	92	93	94	95	96	97	98	99	100

3. Count from 36 to 47. Which of the numbers below will you say? Circle them.

42 31 48 39 37

Tens

Write how many tens. Write the number. (1.NBT.B.2a, 1.NBT.B.2c)

4. _____ tens

_____

5. _____ tens

_____

This page checks understanding of important skills needed for success in Chapter 1.

Name ______________________________

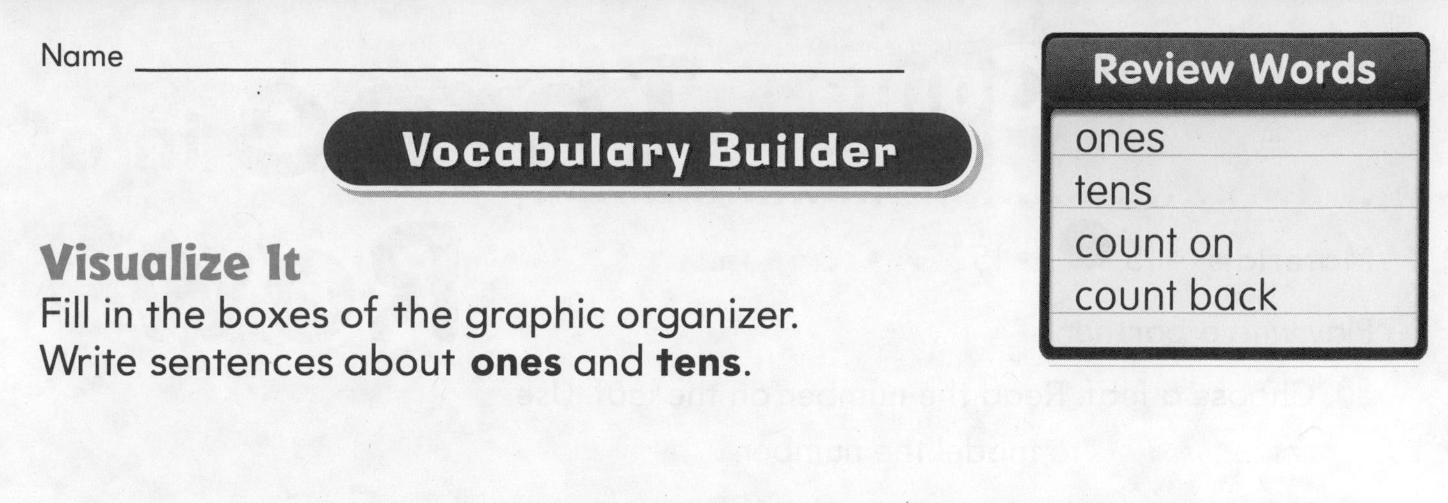

Visualize It

Fill in the boxes of the graphic organizer.
Write sentences about **ones** and **tens**.

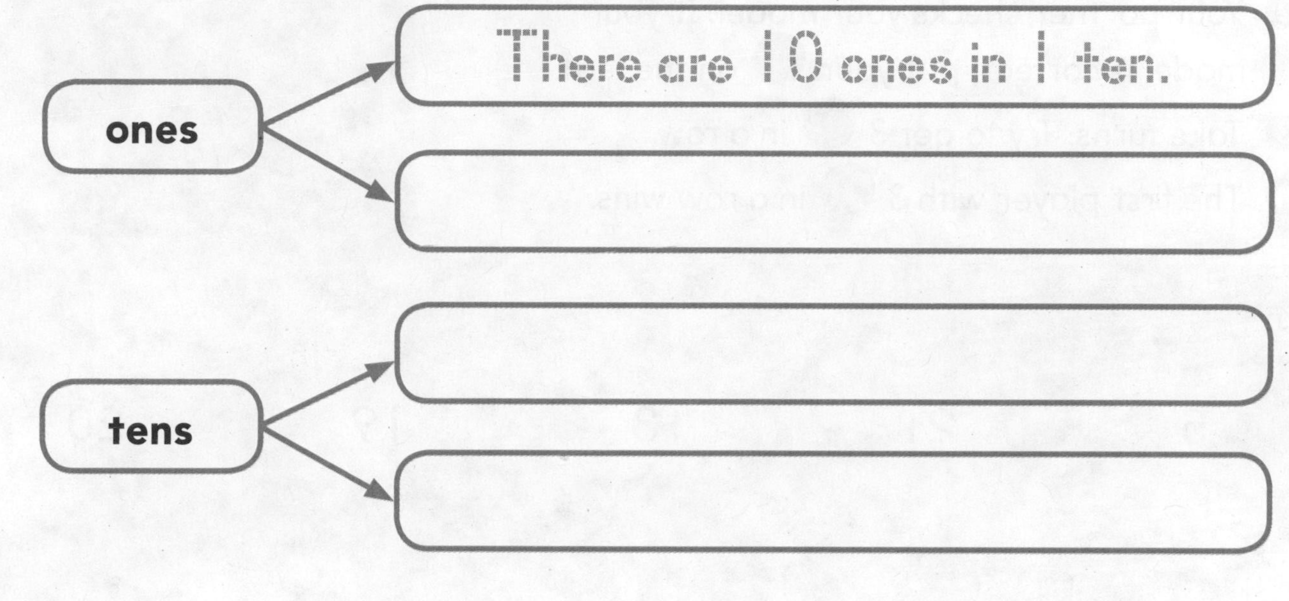

Understand Vocabulary

1. Start with 1. **Count on** by ones.

2. Start with 8. **Count back** by ones.

8, ____, ____, ____, ____, ____

• Interactive Student Edition
• Multimedia eGlossary

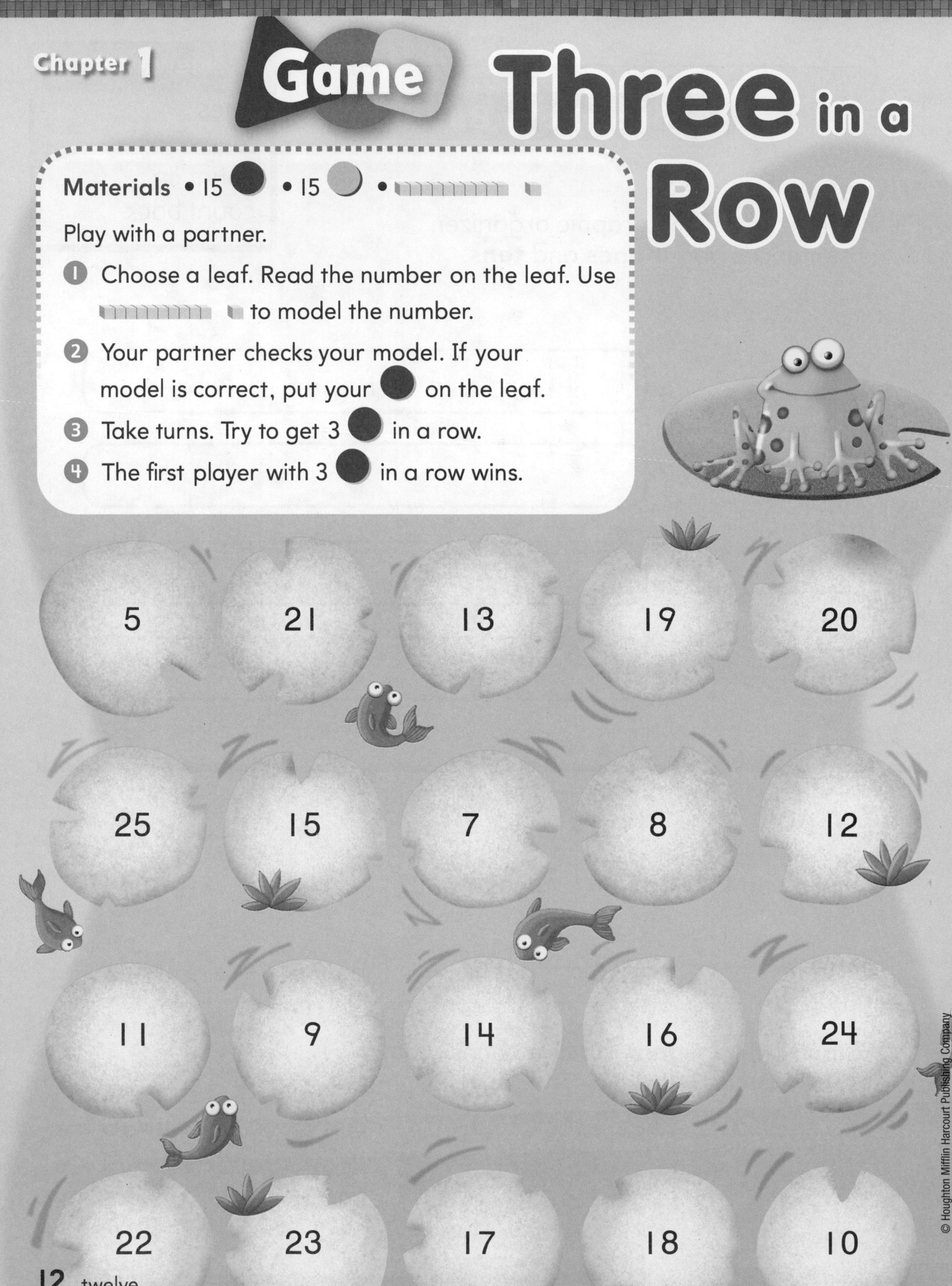

Game: Three in a Row

Materials • 15 ● • 15 ○ • tens rods and ones cubes

Play with a partner.

1. Choose a leaf. Read the number on the leaf. Use tens rods and ones cubes to model the number.
2. Your partner checks your model. If your model is correct, put your ● on the leaf.
3. Take turns. Try to get 3 ● in a row.
4. The first player with 3 ● in a row wins.

5	21	13	19	20
25	15	7	8	12
11	9	14	16	24
22	23	17	18	10

Chapter 1 Vocabulary

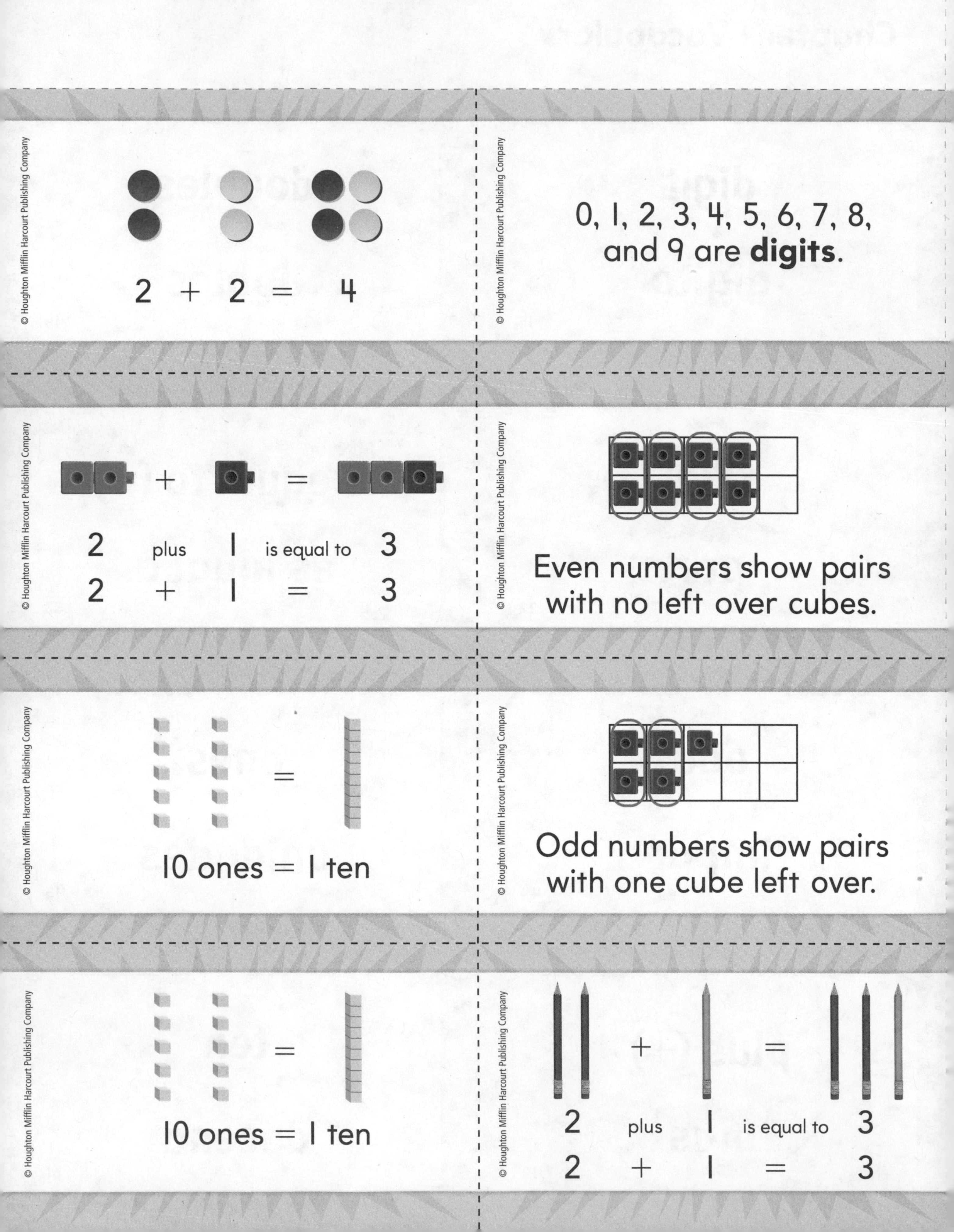

2 + 2 = 4
0, 1, 2, 3, 4, 5, 6, 7, 8, and 9 are **digits**.
2 plus 1 is equal to 3
2 + 1 = 3
Even numbers show pairs with no left over cubes.
=
10 ones = 1 ten
Odd numbers show pairs with one cube left over.
=
10 ones = 1 ten
+ =
2 plus 1 is equal to 3
2 + 1 = 3

Game

Going to the Farmers Market

Word Box

- digit
- doubles
- even numbers
- is equal to (=)
- odd numbers
- ones
- plus (+)
- tens

For 2 to 4 players

Materials

- 1 cube
- 1 cube
- 1 cube
- 1 cube
- 1 number cube

How to Play

1. Put your cube in the START circle of the same color.
2. To get your cube out of START, you must roll a 6.
   - If you do not roll a 6, wait until your next turn.
   - If you roll a 6, move your cube to the circle of that same color on the path.
3. Once you have a cube on the path, toss the number cube to take a turn. Move your cube that many.
4. If you land on a space with a question, answer the question. If your answer is correct, move ahead 1 space.
5. To reach FINISH, you need to move your cube up the path of the same color as the cube. The first player to reach FINISH wins.

Game
START
Which number is equal to 12 + 7?
Which digit is in the ones place in 19?
How can you use doubles to add 4 and 5?
How can you tell when a number is even?
FINISH
Which digit is in the tens place in 45?
How can you tell when a number is odd?
What sign shows that one number is equal to another?
START
How many ones are in 24?

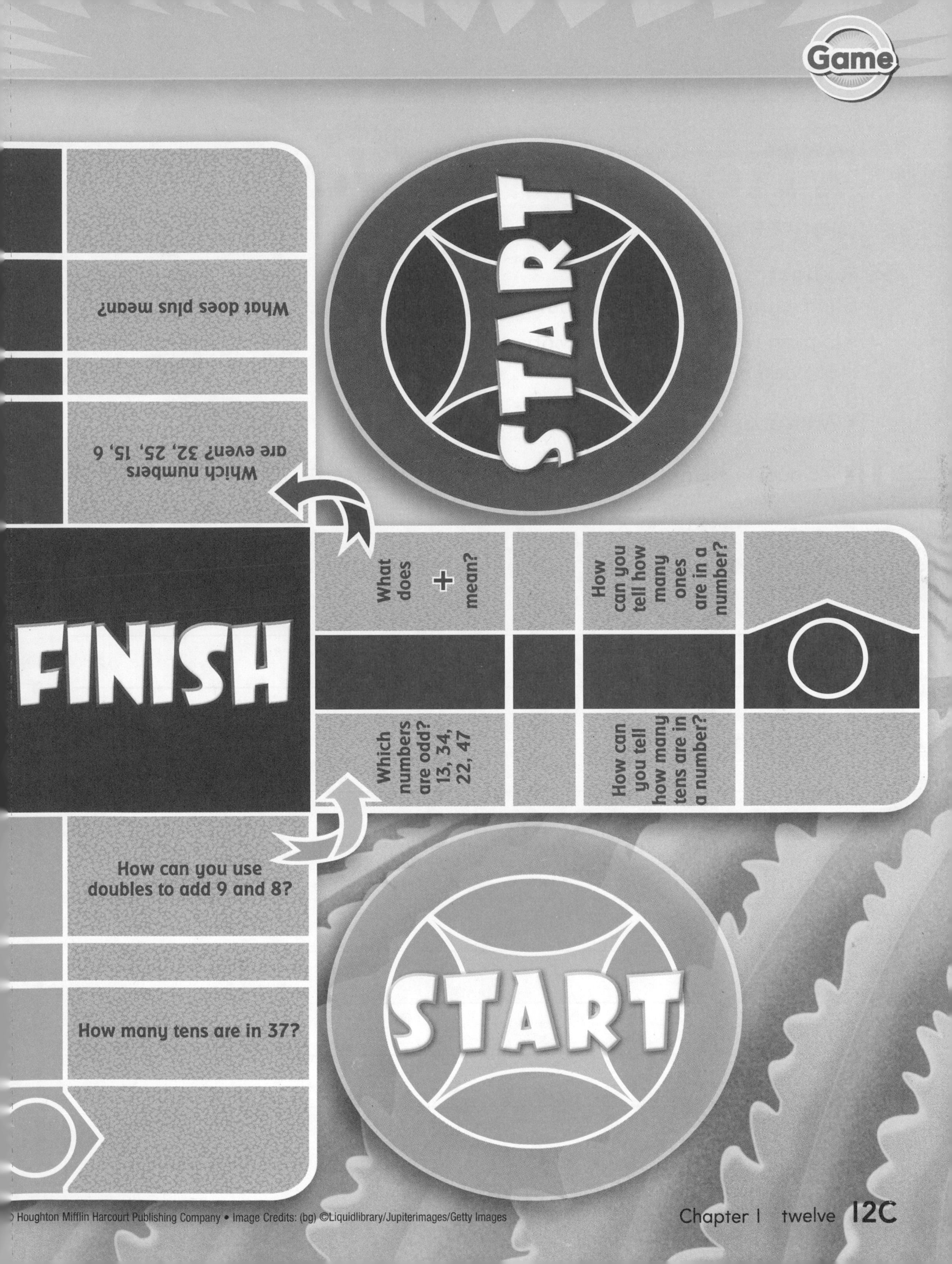
START
What does plus mean?
Which numbers are even? 32, 25, 15, 6
FINISH
What does + mean?
How can you tell how many ones are in a number?
Which numbers are odd? 13, 34, 22, 47
How can you tell how many tens are in a number?
How can you use doubles to add 9 and 8?
How many tens are in 37?
START

The Write Way

Reflect

Choose one idea. Write about it in the space below.

- Explain two things you know about even numbers and odd numbers.
- Write about all the different ways you can show 25.
- Tell how to count on by different amounts to 1,000.

Name ______________________

Algebra • Even and Odd Numbers

Essential Question How are even numbers and odd numbers different?

Common Core **Operations and Algebraic Thinking—2.OA.C.3**

MATHEMATICAL PRACTICES MP3, MP6, MP7

Use ■ to show each number.

FOR THE TEACHER • Read the following problem. Beca has 8 toy cars. Can she put her cars in pairs on a shelf? Have children set pairs of cubes vertically on the ten frames. Continue the activity for the numbers 7 and 10.

Math Talk MATHEMATICAL PRACTICES 6

When you make pairs for 7 and for 10, how are these models different? **Explain.**

Model and Draw

Count out cubes for each number. Make pairs.
Even numbers show pairs with no cubes left over.
Odd numbers show pairs with one cube left over.

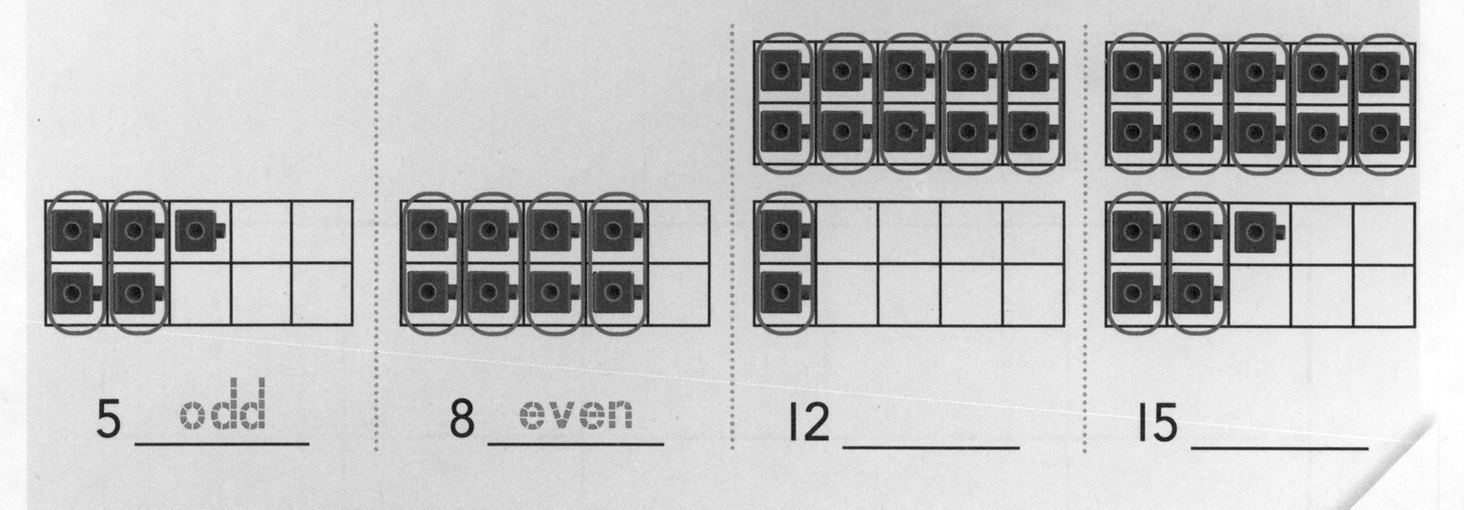

Use cubes. Count out the number of cubes.
Make pairs. Then write **even** or **odd**.

1. 6 ______
2. 3 ______
3. 2 ______
4. 9 ______
5. 4 ______
6. 10 ______
7. 7 ______
8. 13 ______
9. 11 ______
10. 14 ______

Name ___________________________

On Your Own

Shade in the ten frames to show the number. Circle **even** or **odd**.

11. 17

even odd

12. 16

even odd

13. 19

even odd

14. There are an even number of boys and an odd number of girls in Lena's class. How many boys and girls could be in her class? Show your work.

____

____

____

15. MATHEMATICAL PRACTICE 3 **Make Arguments**
Which two numbers in the box are even numbers?

8 5 3 6

________ and ________

Explain how you know that they are even numbers.

____

____

____

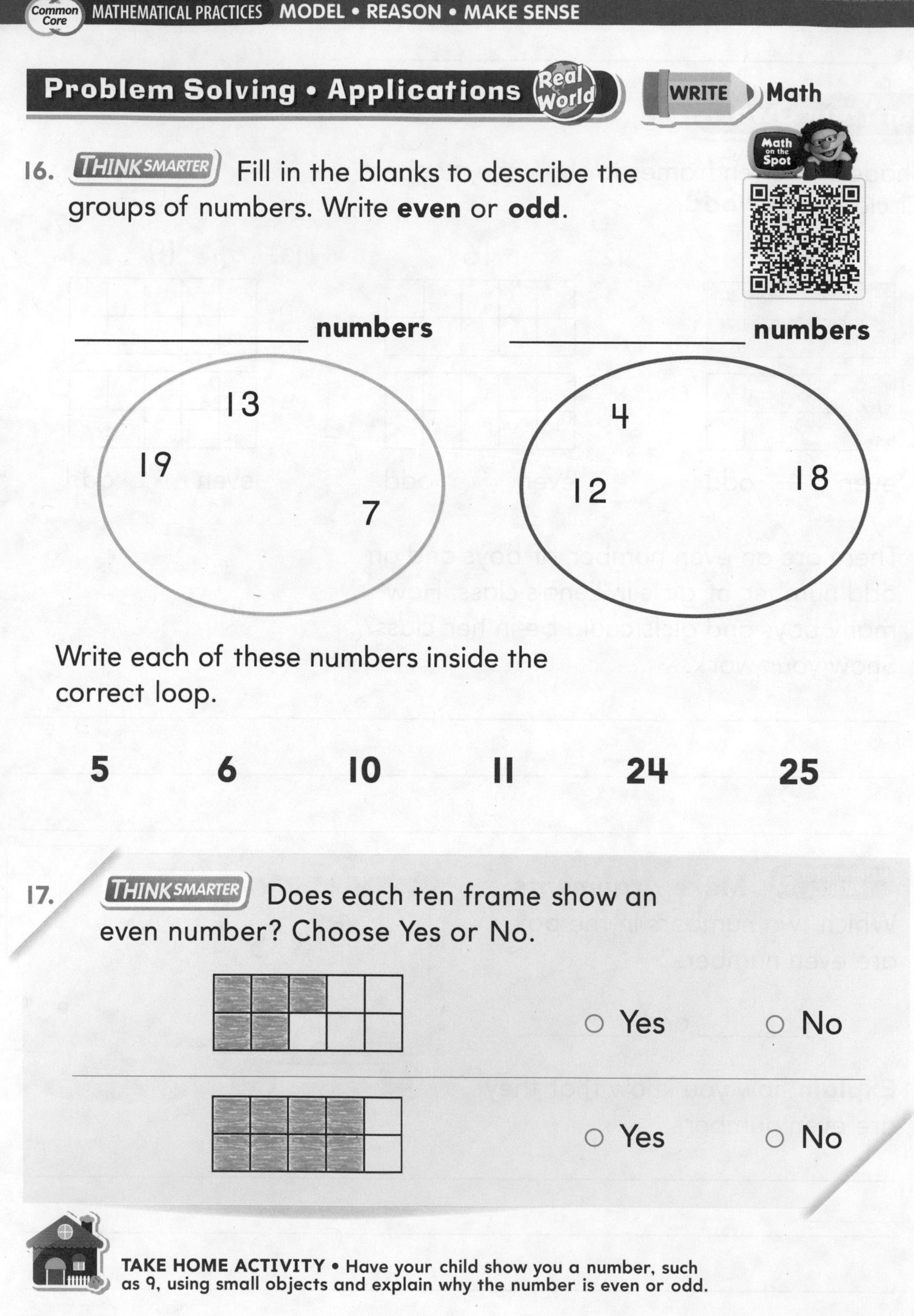

Problem Solving • Applications Real World

WRITE Math

Math on the Spot

16. THINK SMARTER Fill in the blanks to describe the groups of numbers. Write **even** or **odd**.

____________ **numbers** — 13, 19, 7

____________ **numbers** — 4, 12, 18

Write each of these numbers inside the correct loop.

5 6 10 11 24 25

17. THINK SMARTER Does each ten frame show an even number? Choose Yes or No.

○ Yes ○ No

○ Yes ○ No

TAKE HOME ACTIVITY • **Have your child show you a number, such as 9, using small objects and explain why the number is even or odd.**

Name ______________________________

Algebra • Even and Odd Numbers

Common Core **COMMON CORE STANDARD—2.OA.C.3**
Work with equal groups of objects to gain foundations for multiplication.

Shade in the ten frames to show the number. Circle even or odd.

1. 15

even odd

2. 18

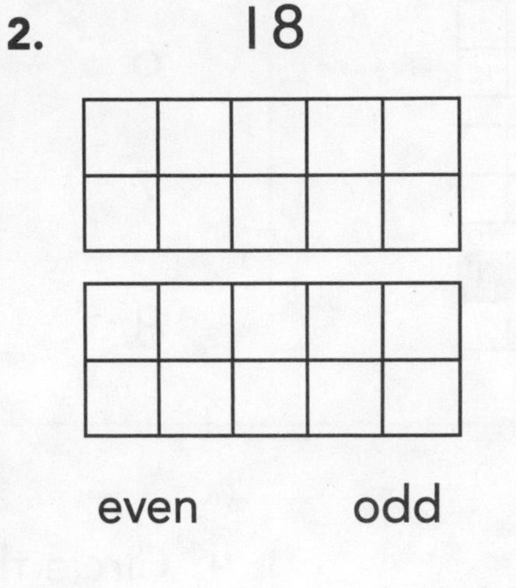

even odd

3. 11

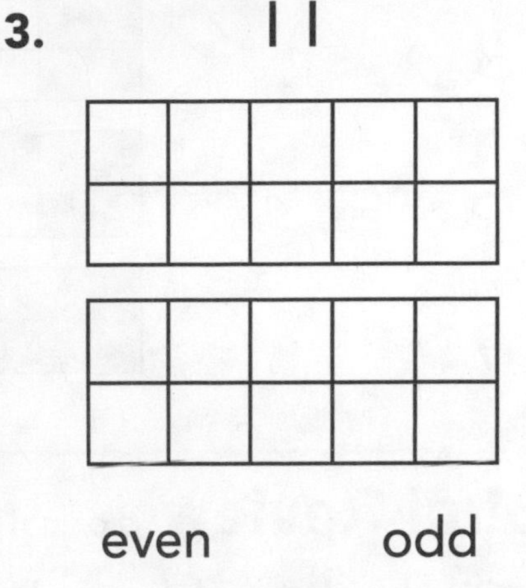

even odd

Problem Solving Real World

4. Mr. Dell has an odd number of sheep and an even number of cows on his farm. Circle the choice that could tell about his farm.

9 sheep and 10 cows

10 sheep and 11 cows

8 sheep and 12 cows

5. WRITE Math Write two odd numbers and two even numbers. Explain how you know which numbers are even and which are odd.

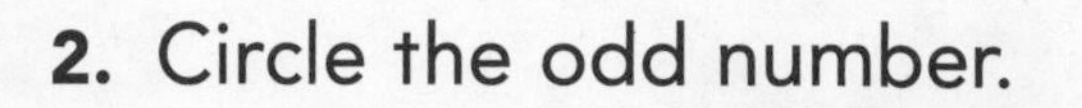

Lesson Check (2.OA.C.3)

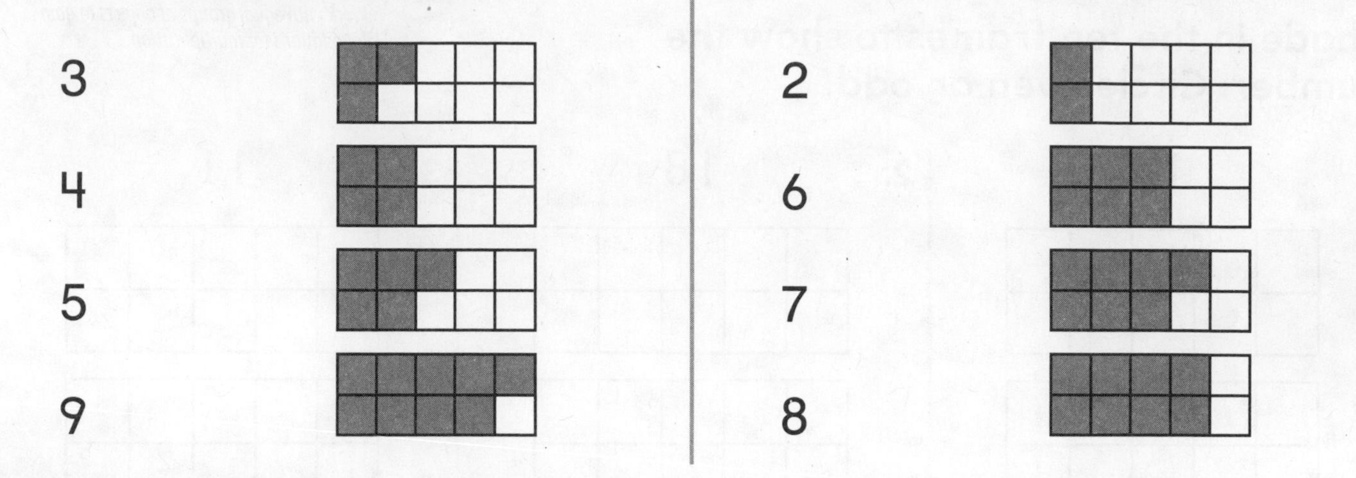

1. Circle the even number.

3
4
5
9

2. Circle the odd number.

2
6
7
8

Spiral Review (2.OA.C.3)

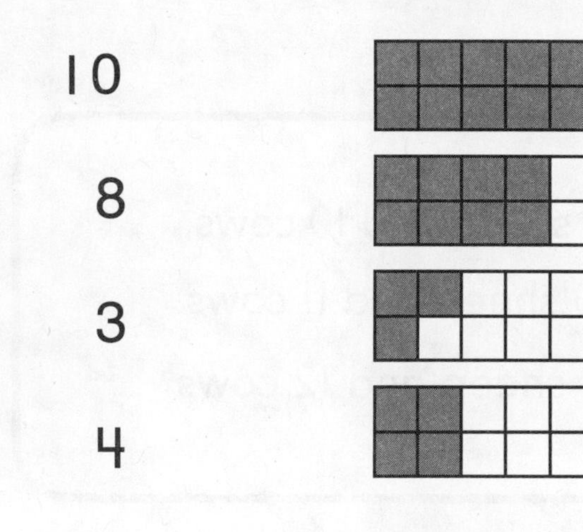

3. Circle the odd number.

10
8
3
4

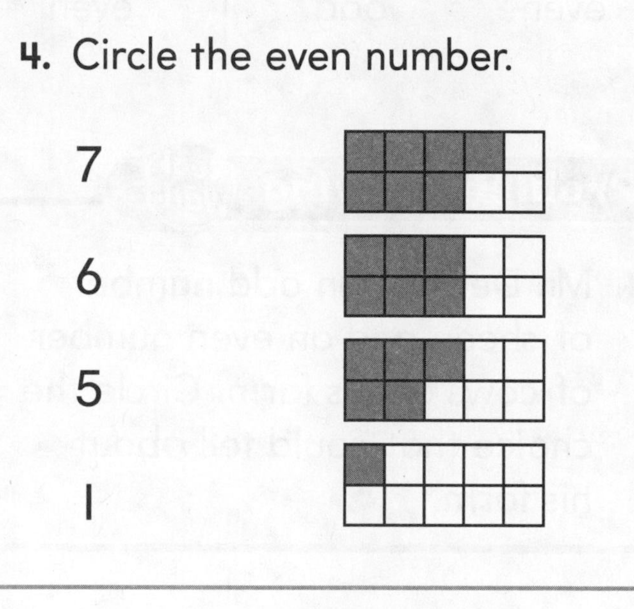

4. Circle the even number.

7
6
5
1

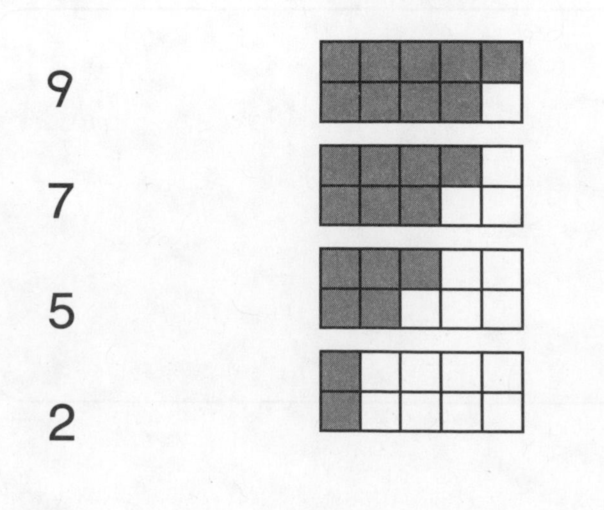

5. Circle the even number.

9
7
5
2

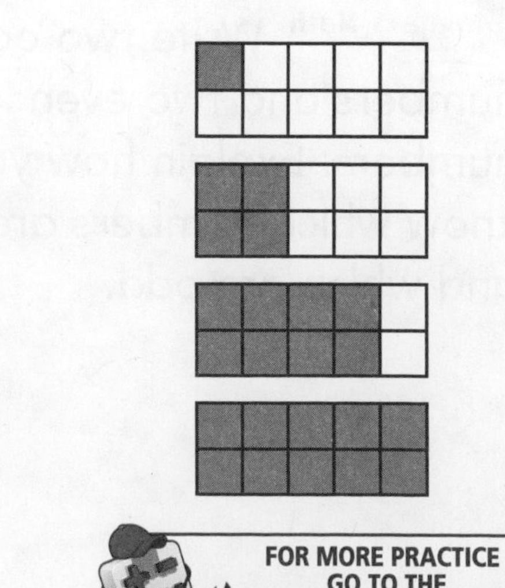

6. Circle the odd number.

1
4
8
10

FOR MORE PRACTICE GO TO THE **Personal Math Trainer**

Name ______________________________

Algebra • Represent Even Numbers

Essential Question Why can an even number be shown as the sum of two equal addends?

Common Core Operations and Algebraic Thinking—2.OA.C.3
MATHEMATICAL PRACTICES
MP2, MP3, MP7, MP8

Listen and Draw

Make pairs with your cubes. Draw to show the cubes. Then write the numbers you say as you count to find the number of cubes.

______________________________ ________ cubes

Math Talk MATHEMATICAL PRACTICES 2

Use Reasoning
Explain how you know if a number modeled with cubes is an even number.

FOR THE TEACHER • Give each small group of children a set of 10 to 15 connecting cubes. After children group their cubes into pairs, have them draw a picture of their cubes and write their counting sequence for finding the total number of cubes.

Model and Draw

An even number of cubes can be shown as two equal groups.

You can match each cube in the first group with a cube in the second group.

$6 = 3 + 3$

$10 = 5 + 5$

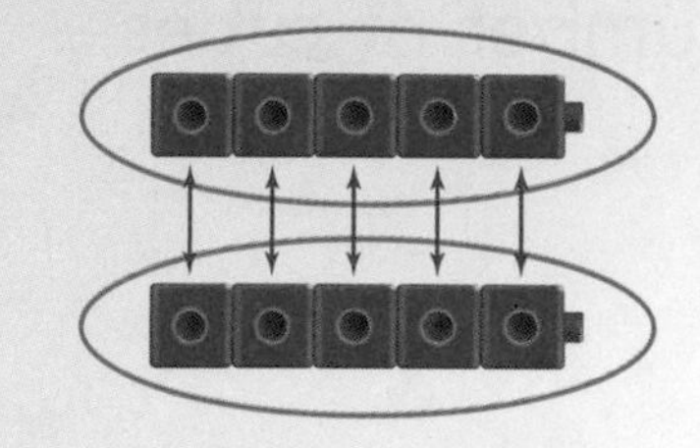

Share and Show

How many cubes are there in all? Complete the addition sentence to show the equal groups.

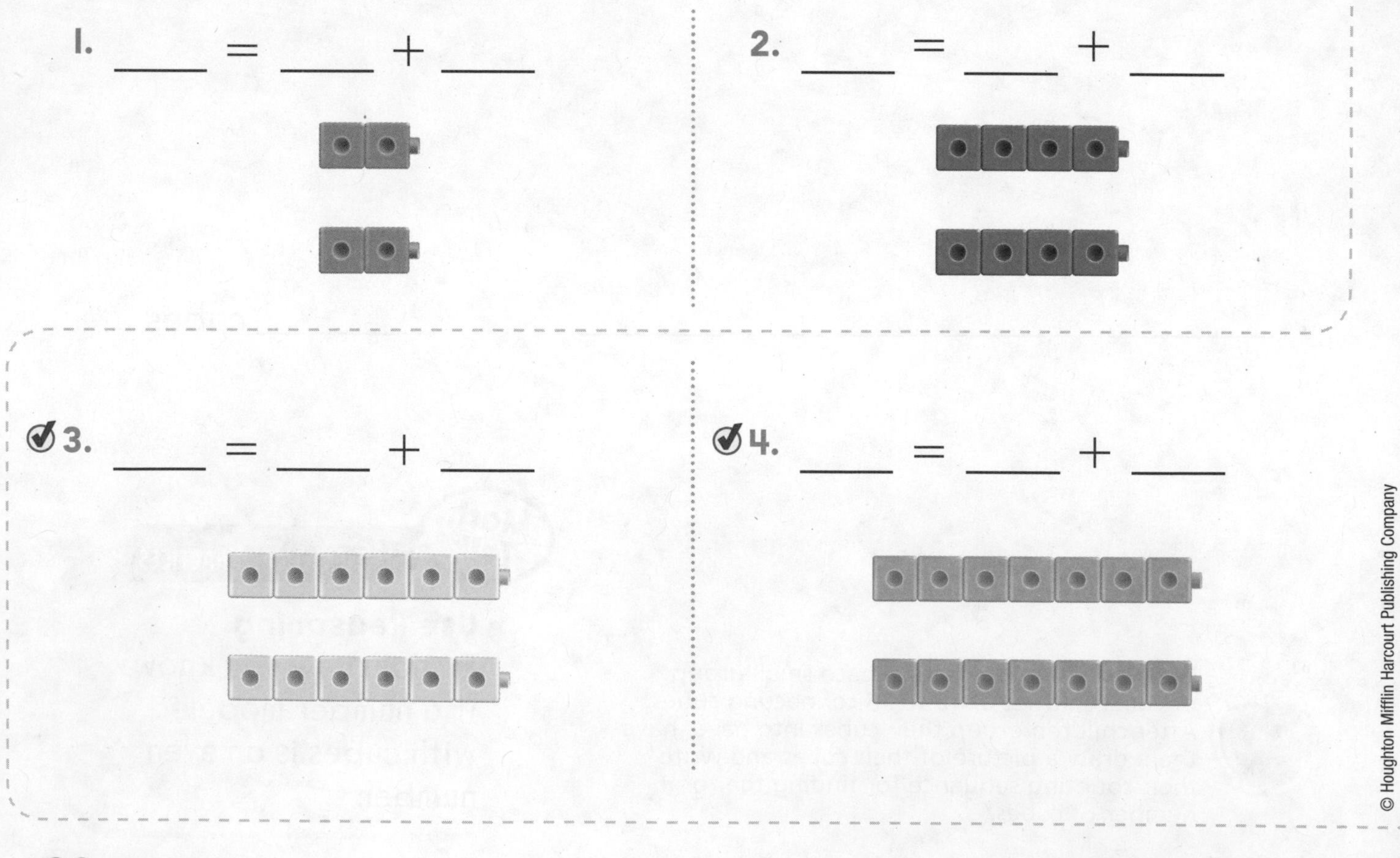

1. ____ = ____ + ____

2. ____ = ____ + ____

✓3. ____ = ____ + ____

✓4. ____ = ____ + ____

Name ______________________________

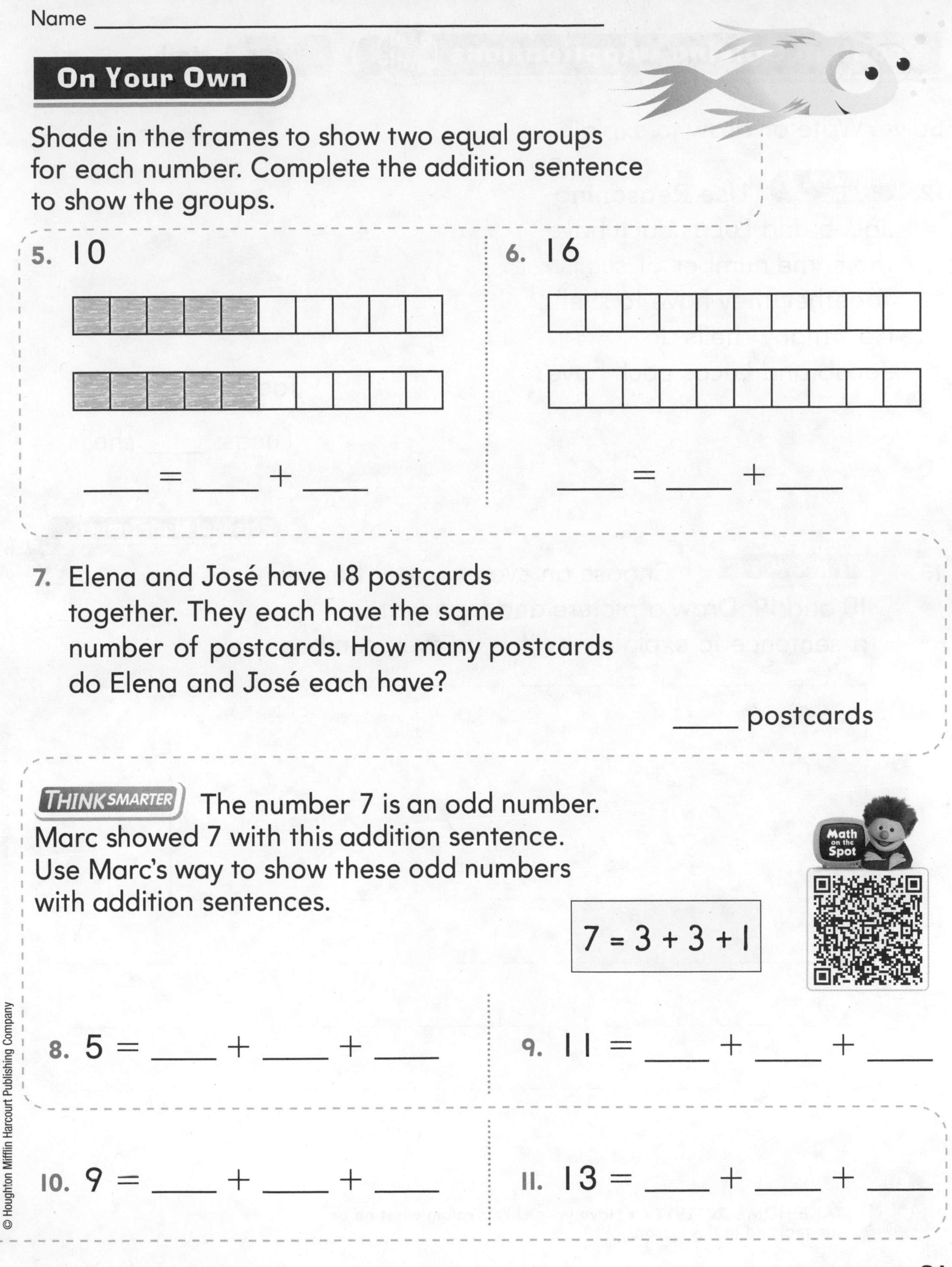

On Your Own

Shade in the frames to show two equal groups for each number. Complete the addition sentence to show the groups.

5. 10

____ = ____ + ____

6. 16

____ = ____ + ____

7. Elena and José have 18 postcards together. They each have the same number of postcards. How many postcards do Elena and José each have?

____ postcards

THINK SMARTER The number 7 is an odd number. Marc showed 7 with this addition sentence. Use Marc's way to show these odd numbers with addition sentences.

7 = 3 + 3 + 1

Math on the Spot

8. 5 = ____ + ____ + ____

9. 11 = ____ + ____ + ____

10. 9 = ____ + ____ + ____

11. 13 = ____ + ____ + ____

Problem Solving • Applications

Real World

WRITE Math

Solve. Write or draw to explain.

12. MATHEMATICAL PRACTICE 2 **Use Reasoning**
Jacob and Lucas each have the same number of shells. Together they have 16 shells. How many shells do Jacob and Lucas each have?

Jacob: _____ shells

Lucas: _____ shells

Personal Math Trainer

13. THINKSMARTER+ Choose an even number between 10 and 19. Draw a picture and then write a sentence to explain why it is an even number.

____

____

____

TAKE HOME ACTIVITY • Have your child explain what he or she learned in this lesson.

Name ______________________

Algebra • Represent Even Numbers

Common Core **COMMON CORE STANDARD—2.OA.C.3**
Work with equal groups of objects to gain foundations for multiplication.

Shade in the frames to show two equal groups for each number. Complete the addition sentence to show the groups.

1. 8

____ = ____ + ____

2. 18

____ = ____ + ____

3. 10

____ = ____ + ____

4. 14

____ = ____ + ____

5. 20

____ = ____ + ____

Problem Solving Real World

Solve. Write or draw to explain.

6. The seats in a van are in pairs. There are 16 seats. How many pairs of seats are there?

____ pairs of seats

7. WRITE Math Draw or write to show that the number 18 is an even number.

Lesson Check (2.OA.C.3)

1. Circle the sum that is an even number.

9 + 9 = 18
9 + 8 = 17
8 + 7 = 15
6 + 5 = 11

2. Circle the sum that is an even number.

1 + 2 = 3
3 + 3 = 6
2 + 5 = 7
4 + 7 = 11

Spiral Review (2.OA.C.3)

3. Circle the even number.

7
9
10
13

4. Circle the odd number.

4
11
16
20

5. Ray has an odd number of cats. He also has an even number of dogs. Complete the sentence.

Ray has _____ cats and _____ dogs.

6. Circle the sum that is an even number.

2 + 3 = 5
3 + 4 = 7
4 + 4 = 8
7 + 8 = 15

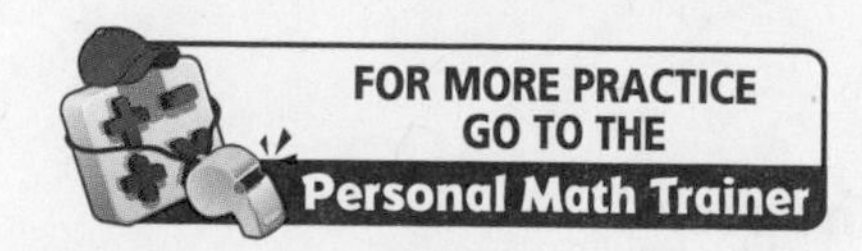

Name ______________________________

Understand Place Value

Essential Question How do you know the value of a digit?

Common Core **Number and Operations in Base Ten—2.NBT.A.3**

MATHEMATICAL PRACTICES MP1, MP6

Write the numbers. Then choose a way to show the numbers.

Tens	Ones

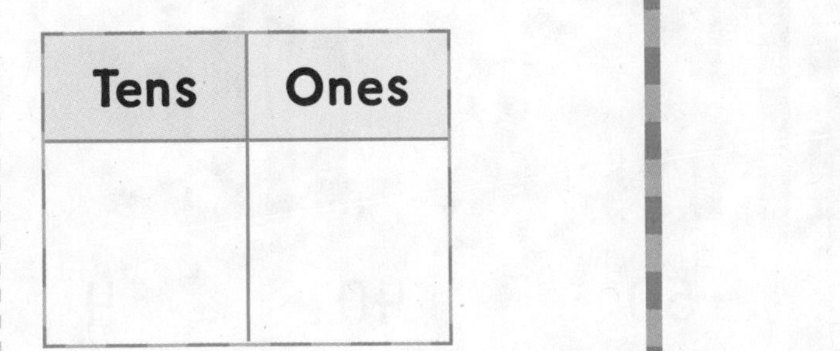

Tens	Ones

FOR THE TEACHER • Read the following problem. Have children write the numbers and describe how they chose to represent them. Gabriel collects baseball cards. The number of cards that he has is written with a 2 and a 5. How many cards might he have?

Math Talk MATHEMATICAL PRACTICES 6

Explain why the value of 5 is different in the two numbers.

Model and Draw

0, 1, 2, 3, 4, 5, 6, 7, 8, and 9 are **digits**.
In a 2-digit number, you know the value of a digit by its place.

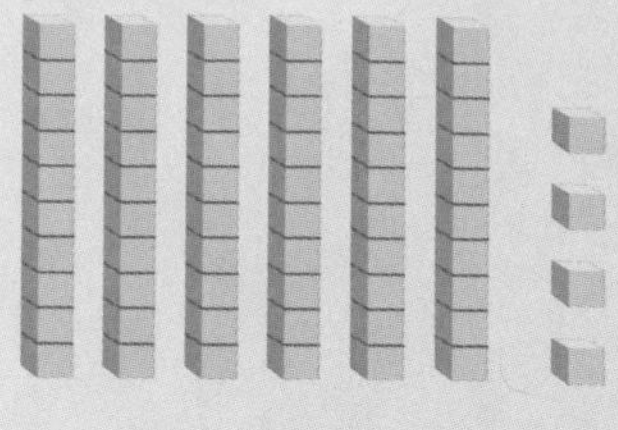

64

Tens	Ones
6	4

6 tens   4 ones

The digit **6** is in the tens place. It tells you there are 6 tens, or 60.

The digit **4** is in the ones place. It tells you there are 4 ones, or 4.

Share and Show

MATH BOARD

Circle the value of the red digit.

1. 26
   60   6

2. 58
   5   50

3. 40
   40   4

4. 73
   30   3

5. 24
   2   20

6. 61
   1   10

Name ______________________________

On Your Own

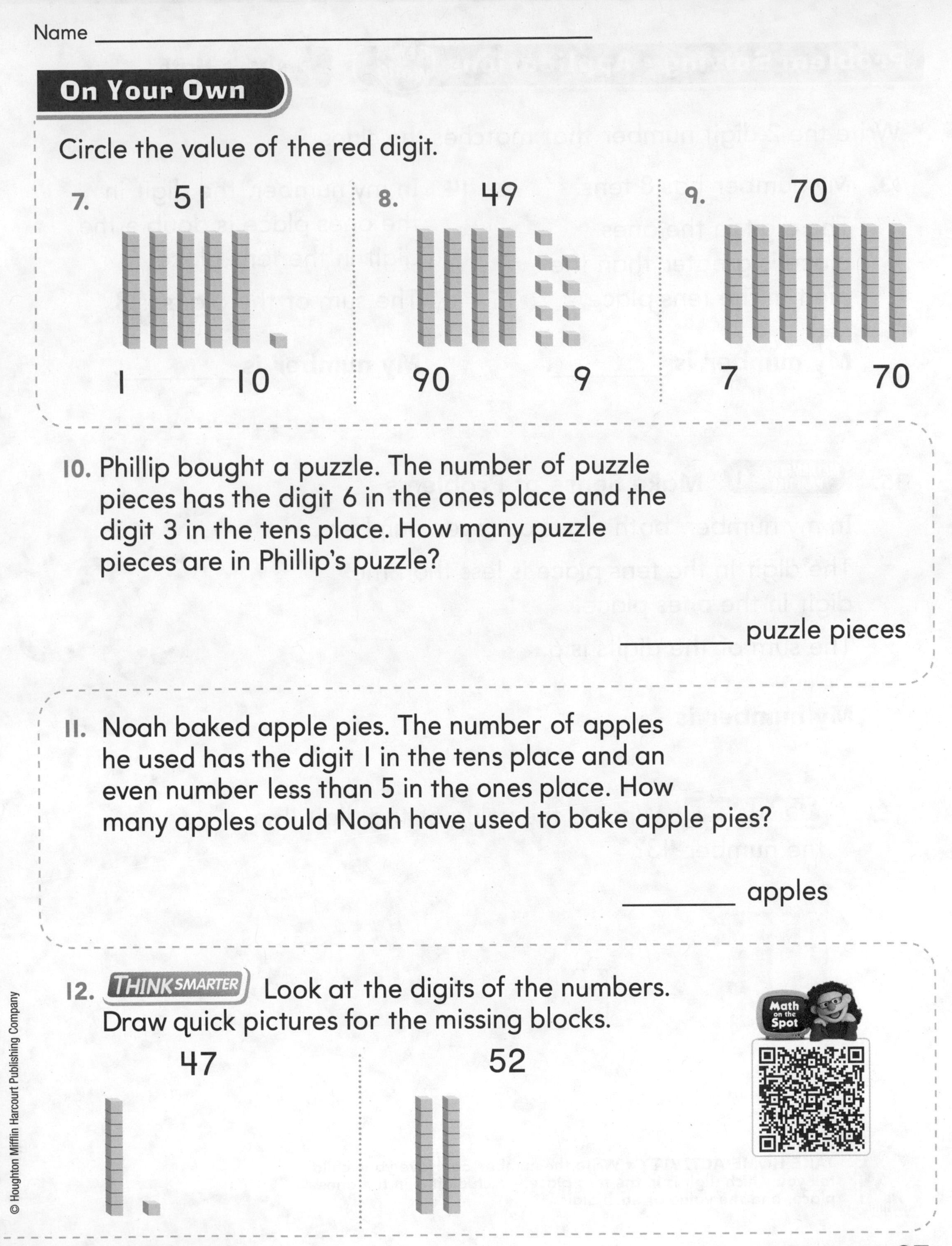

Circle the value of the red digit.

7. 51

1 10

8. 49

90 9

9. 70

7 70

10. Phillip bought a puzzle. The number of puzzle pieces has the digit 6 in the ones place and the digit 3 in the tens place. How many puzzle pieces are in Phillip's puzzle?

_______ puzzle pieces

11. Noah baked apple pies. The number of apples he used has the digit 1 in the tens place and an even number less than 5 in the ones place. How many apples could Noah have used to bake apple pies?

_______ apples

12. THINKSMARTER Look at the digits of the numbers. Draw quick pictures for the missing blocks.

47

52

Problem Solving • Applications

Write the 2-digit number that matches the clues.

13. My number has 8 tens.

    The digit in the ones place is greater than the digit in the tens place.

    My number is ________.

14. In my number, the digit in the ones place is double the digit in the tens place.

    The sum of the digits is 3.

    My number is ________.

15. 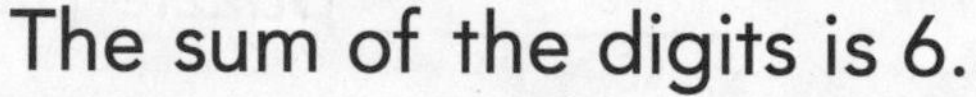

    In my number, both digits are even numbers.

    The digit in the tens place is less than the digit in the ones place.

    The sum of the digits is 6.

    My number is ________.

16. THINKSMARTER What is the value of the digit 4 in the number 43?

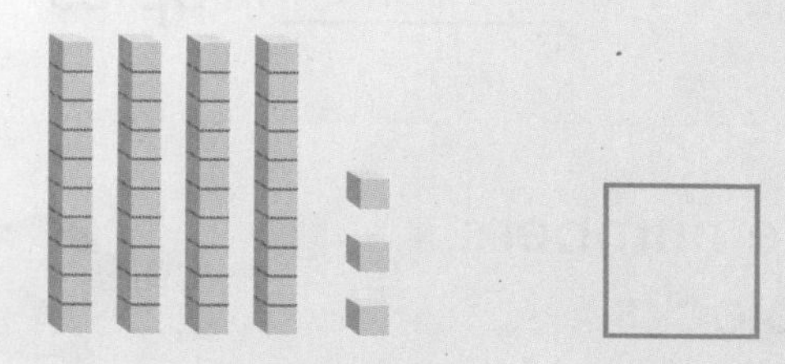

TAKE HOME ACTIVITY • Write the number 56. Have your child tell you which digit is in the tens place, which digit is in the ones place, and the value of each digit.

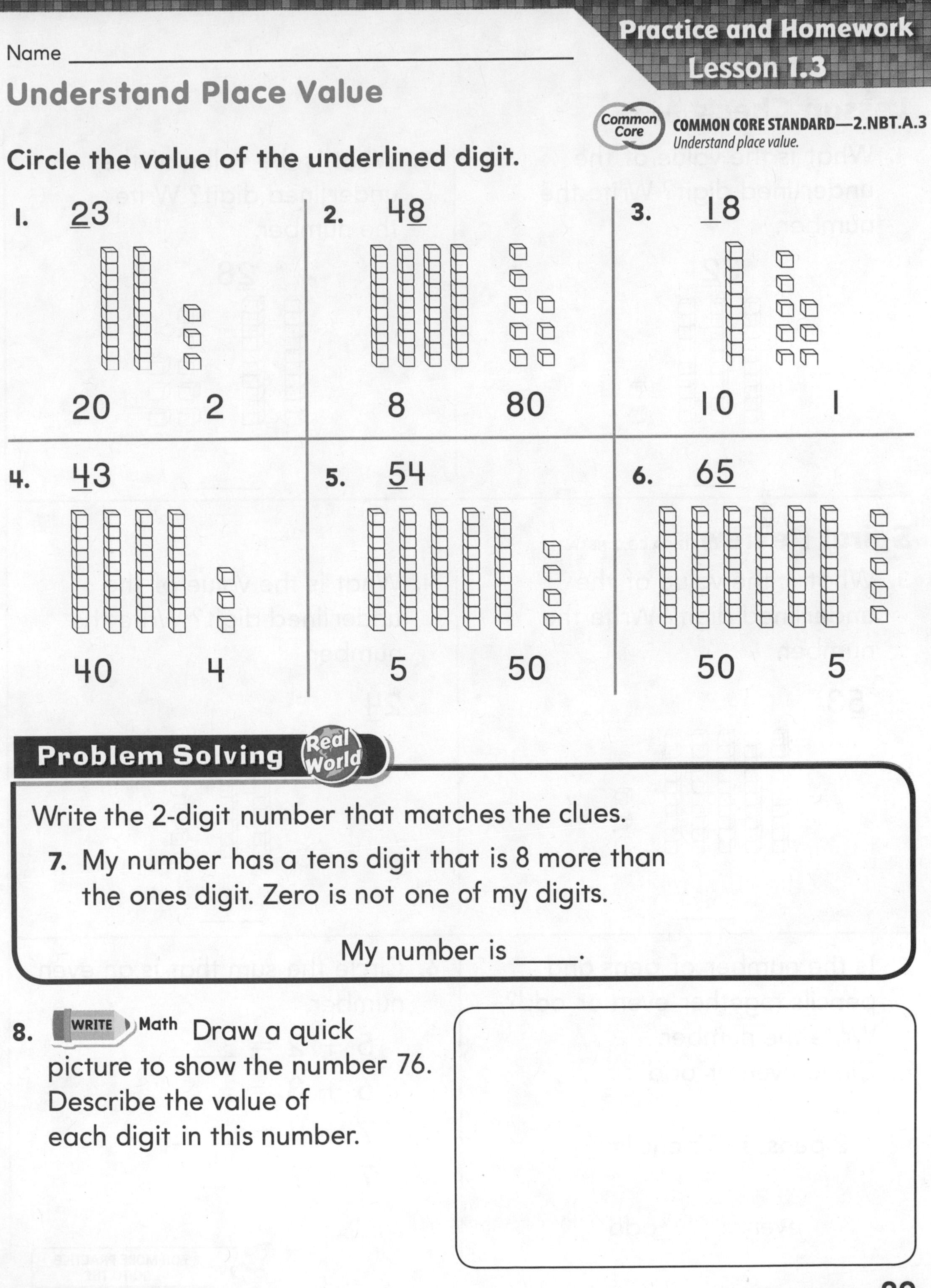

Name ____________________

Understand Place Value

Common Core — COMMON CORE STANDARD—2.NBT.A.3
Understand place value.

Circle the value of the underlined digit.

1. <u>2</u>3

20 2

2. 4<u>8</u>

8 80

3. <u>1</u>8

10 1

4. <u>4</u>3

40 4

5. <u>5</u>4

5 50

6. 6<u>5</u>

50 5

Problem Solving — Real World

Write the 2-digit number that matches the clues.

7. My number has a tens digit that is 8 more than the ones digit. Zero is not one of my digits.

My number is ____.

8. WRITE Math Draw a quick picture to show the number 76. Describe the value of each digit in this number.

Lesson Check (2.NBT.A.3)

1. What is the value of the underlined digit? Write the number.

3<u>2</u>

____

2. What is the value of the underlined digit? Write the number.

<u>2</u>8

____

Spiral Review (2.OA.C.3, 2.NBT.A.3)

3. What is the value of the underlined digit? Write the number.

<u>5</u>3

____

4. What is the value of the underlined digit? Write the number.

2<u>4</u>

____

5. Is the number of pens and pencils together even or odd? Write the number. Circle even or odd.

2 pens + 3 pencils ____

even odd

6. Circle the sum that is an even number.

$5 + 2 =$ ____

$6 + 3 =$ ____

$7 + 4 =$ ____

$7 + 7 =$ ____

FOR MORE PRACTICE GO TO THE Personal Math Trainer

Name ______________________________

Expanded Form

Essential Question How do you describe a 2-digit number as tens and ones?

Common Core **Number and Operations in Base Ten—2.NBT.A.3**

MATHEMATICAL PRACTICES
MP4, MP6

Use [base-ten blocks] or *i*Tools to model each number.

Tens	Ones

Math Talk — MATHEMATICAL PRACTICES 6

Explain how you know how many tens and ones are in the number 29.

FOR THE TEACHER • After you read the following problem, write 38 on the board. Have children model the number. Emmanuel put 38 stickers on his paper. How can you model 38 with blocks? Continue the activity for 83 and 77.

Model and Draw

What does 23 mean?

Tens	Ones

The 2 in 23 has a value of 2 tens, or 20.
The 3 in 23 has a value of 3 ones, or 3.

2 tens 3 ones

20 + 3

Draw a quick picture to show the number.
Describe the number in two ways.

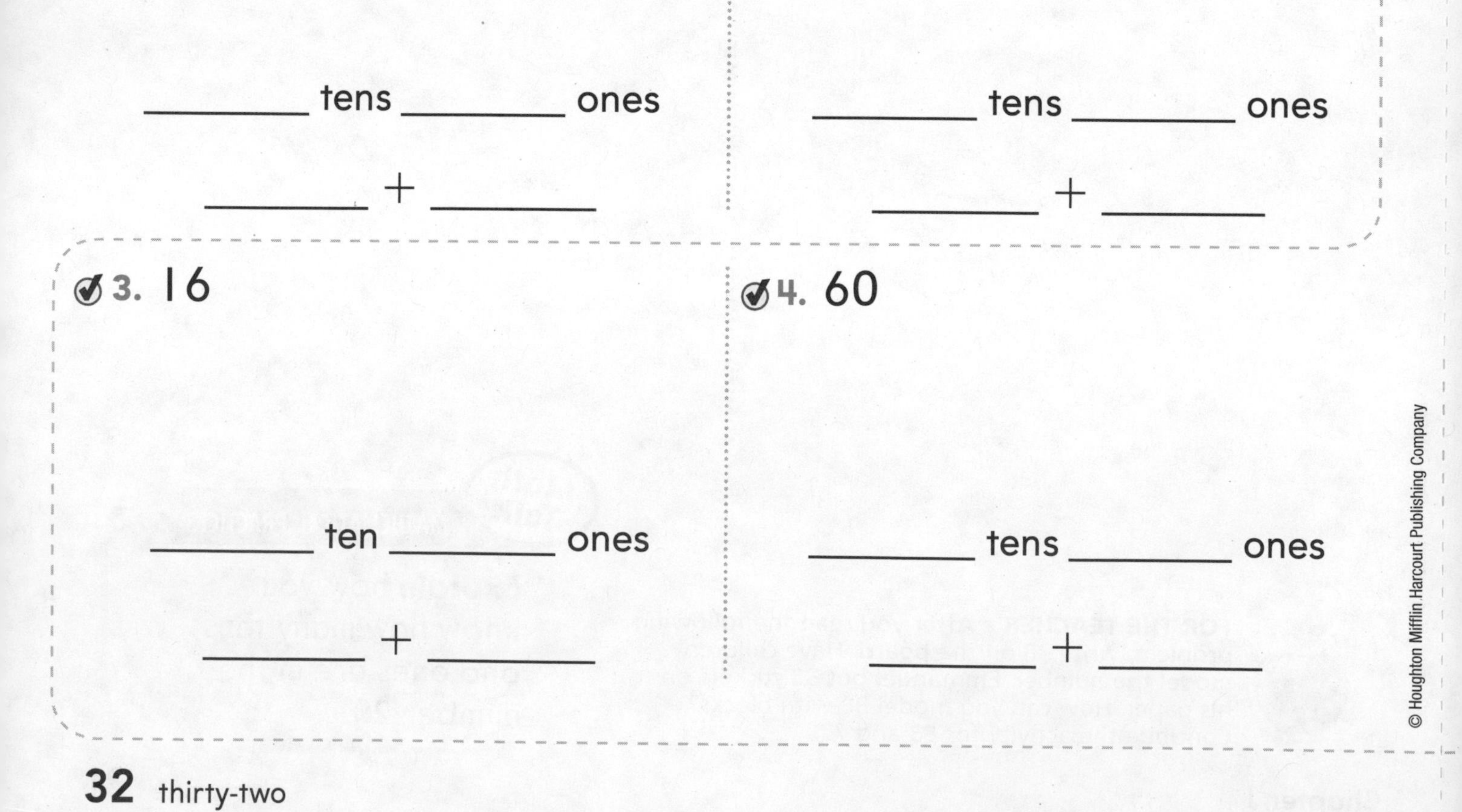

1. 37

_____ tens _____ ones

_____ + _____

2. 54

_____ tens _____ ones

_____ + _____

✓3. 16

_____ ten _____ ones

_____ + _____

✓4. 60

_____ tens _____ ones

_____ + _____

Name ______________________________

On Your Own

Draw a quick picture to show the number. Describe the number in two ways.

5. 48

_______ tens _______ ones

_______ + _______

6. 31

_______ tens _______ one

_______ + _______

7. Riley has some toy dinosaurs. The number she has is one less than 50. Describe the number of toy dinosaurs in two ways.

____

____

Solve. Write or draw to explain.

8. THINK SMARTER Eric has 4 bags of 10 marbles and 6 single marbles. How many marbles does Eric have?

_______ marbles

Problem Solving • Applications

Real World

WRITE Math

MATHEMATICAL PRACTICE 6 **Make Connections**

Use crayons. Follow the steps.

9. Start at 51 and draw a green line to 43.
10. Draw a blue line from 43 to 34.
11. Draw a red line from 34 to 29.
12. Then draw a yellow line from 29 to 72.

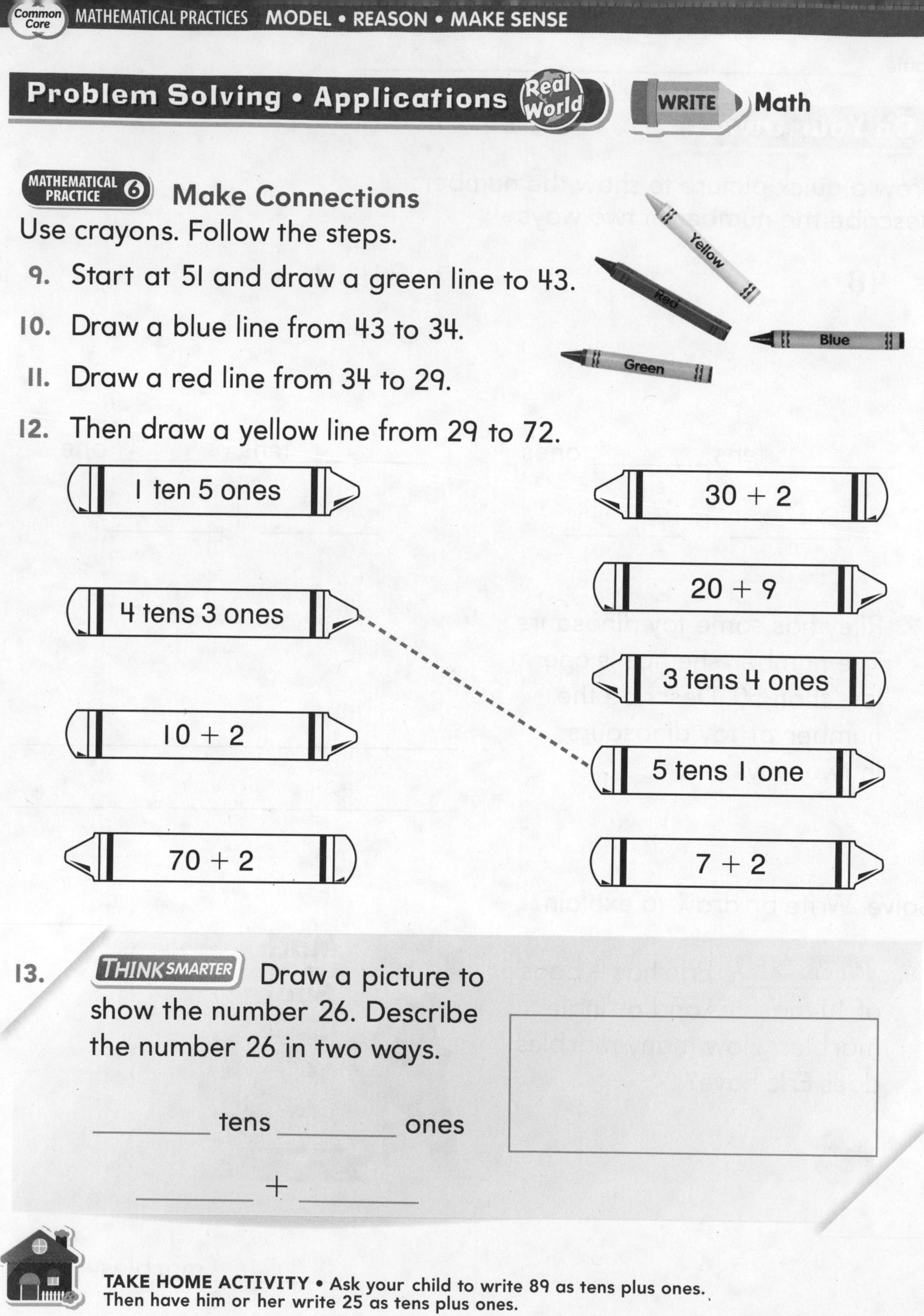

13. THINK SMARTER Draw a picture to show the number 26. Describe the number 26 in two ways.

_______ tens _______ ones

_______ + _______

TAKE HOME ACTIVITY • Ask your child to write 89 as tens plus ones. Then have him or her write 25 as tens plus ones.

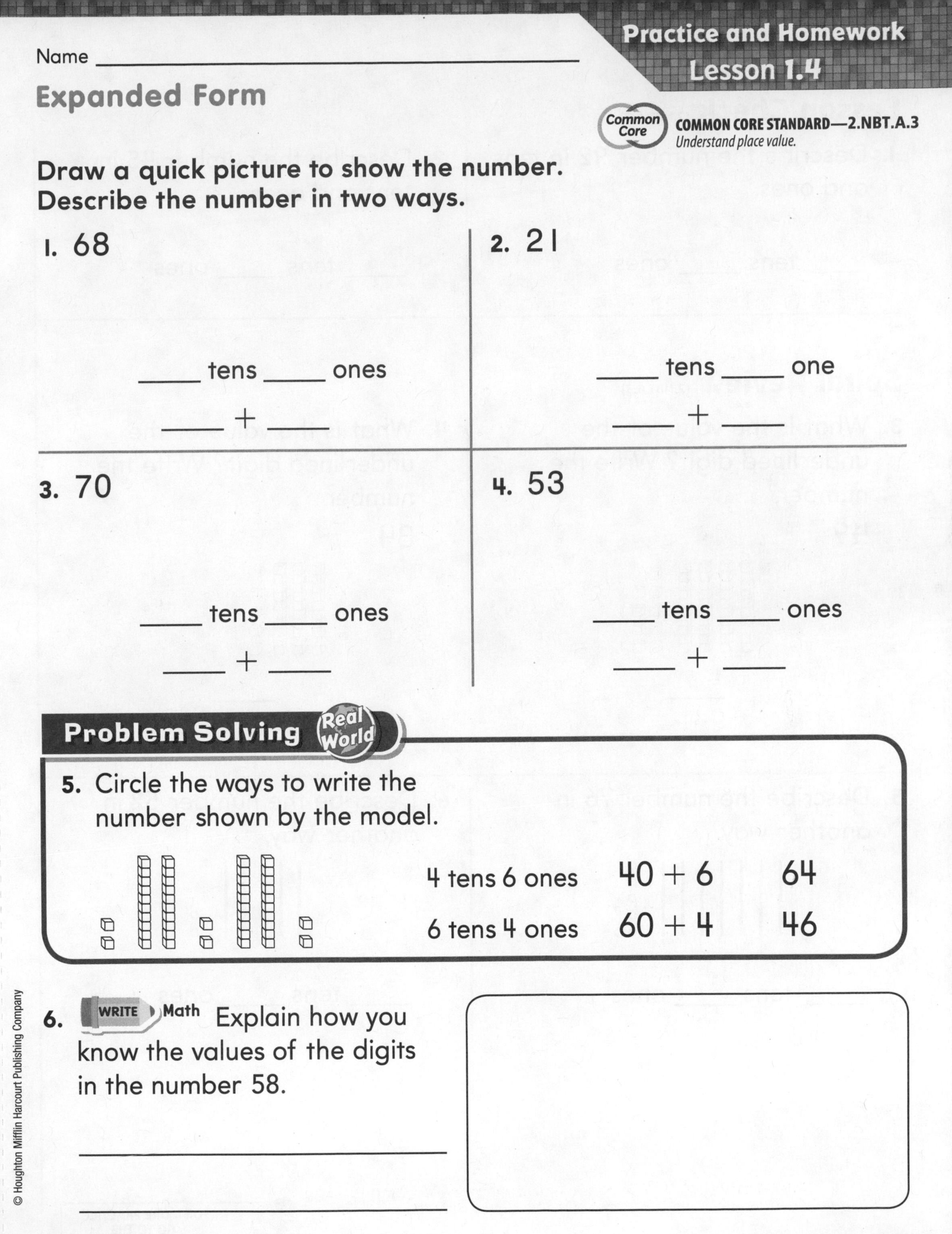

Name ______________________________

Expanded Form

COMMON CORE STANDARD—2.NBT.A.3
Understand place value.

Draw a quick picture to show the number.
Describe the number in two ways.

1. 68

____ tens ____ ones

____ + ____

2. 21

____ tens ____ one

____ + ____

3. 70

____ tens ____ ones

____ + ____

4. 53

____ tens ____ ones

____ + ____

Problem Solving Real World

5. Circle the ways to write the number shown by the model.

4 tens 6 ones   40 + 6   64

6 tens 4 ones   60 + 4   46

6. WRITE Math Explain how you know the values of the digits in the number 58.

____

____

Lesson Check (2.NBT.A.3)

1. Describe the number 92 in tens and ones.

_____ tens _____ ones

2. Describe the number 45 in tens and ones.

_____ tens _____ ones

Spiral Review (2.NBT.A.3)

3. What is the value of the underlined digit? Write the number.

49

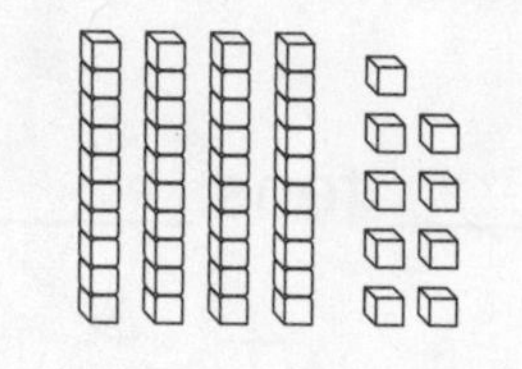

_____

4. What is the value of the underlined digit? Write the number.

34

_____

5. Describe the number 76 in another way.

_____ tens _____ ones

6. Describe the number 52 in another way.

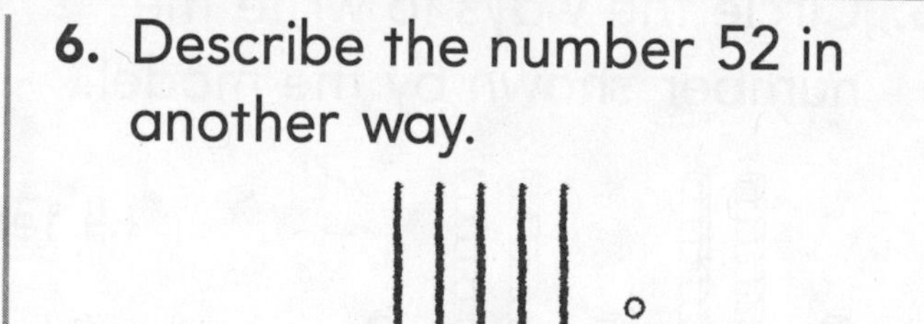

_____ tens _____ ones

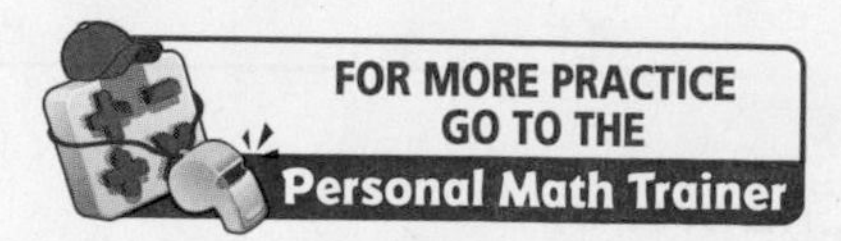

Name ____________________

Different Ways to Write Numbers

Essential Question What are different ways to write a 2-digit number?

Common Core **Number and Operations in Base Ten—2.NBT.A.3**

MATHEMATICAL PRACTICES
MP1, MP6

Listen and Draw Real World

Write the number.
Then write it as tens and ones.

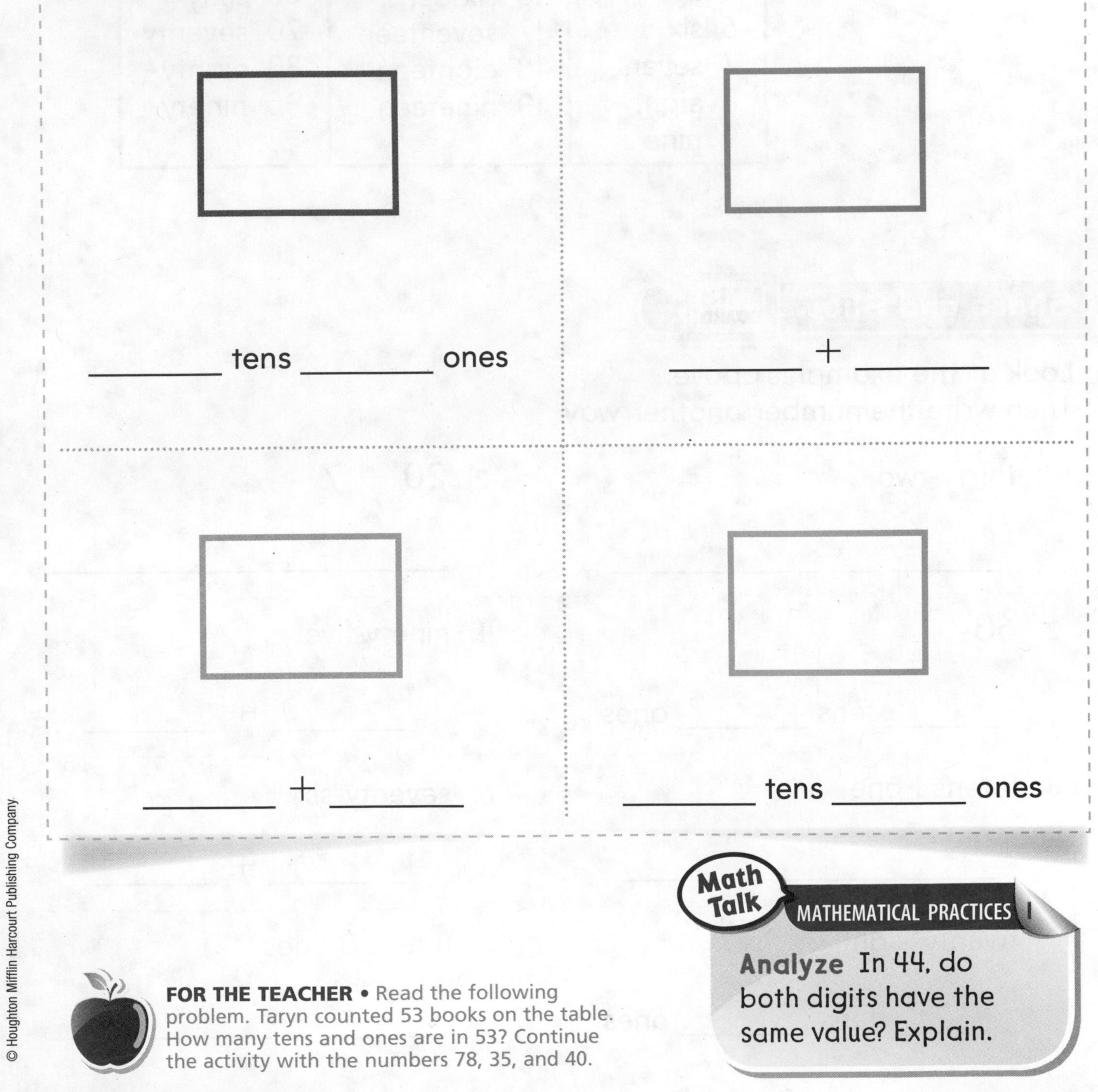

Math Talk **MATHEMATICAL PRACTICES** 1

Analyze In 44, do both digits have the same value? Explain.

FOR THE TEACHER • Read the following problem. Taryn counted 53 books on the table. How many tens and ones are in 53? Continue the activity with the numbers 78, 35, and 40.

Model and Draw

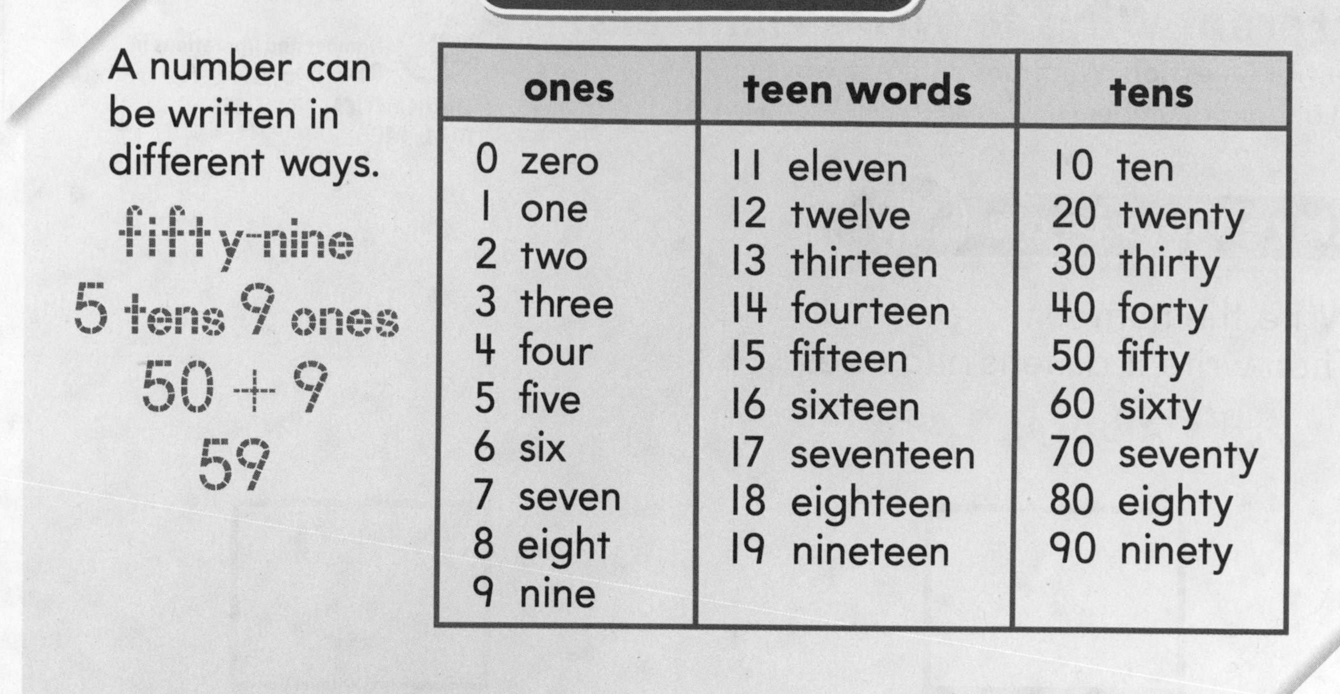

A number can be written in different ways.

fifty-nine

5 tens 9 ones

50 + 9

59

ones	teen words	tens
0 zero	11 eleven	10 ten
1 one	12 twelve	20 twenty
2 two	13 thirteen	30 thirty
3 three	14 fourteen	40 forty
4 four	15 fifteen	50 fifty
5 five	16 sixteen	60 sixty
6 six	17 seventeen	70 seventy
7 seven	18 eighteen	80 eighty
8 eight	19 nineteen	90 ninety
9 nine		

Share and Show

Look at the examples above.
Then write the number another way.

1. thirty-two

____

2. 20 + 7

____

3. 63

_______ tens _______ ones

4. ninety-five

_______ + _______

5. 5 tens 1 one

____

6. seventy-six

_______ + _______

✓7. twenty-eight

_______ tens _______ ones

✓8. 8 tens 0 ones

____

Name ______________________

On Your Own

Write the number another way.

9. 2 tens 4 ones

____

10. thirty

_______ tens _______ ones

11. eighty-five

____

12. 54

_______ + _______

13. Lee has a favorite number. The number has the digit 3 in the ones place and the digit 9 in the tens place. What is another way to write this number?

____

14. Dan's number has a digit greater than 5 in the ones place and a digit less than 5 in the tens place. What could be Dan's number?

____

THINKSMARTER Fill in the blanks to make the sentence true.

Math on the Spot

15. Sixty-seven is the same as _______ tens _______ ones.

16. 4 tens _______ ones is the same as _______ + _______.

17. 20 + _______ is the same as ______________.

TAKE HOME ACTIVITY • Write 20 + 6 on a sheet of paper. Have your child write the 2-digit number. Repeat for 4 tens 9 ones.

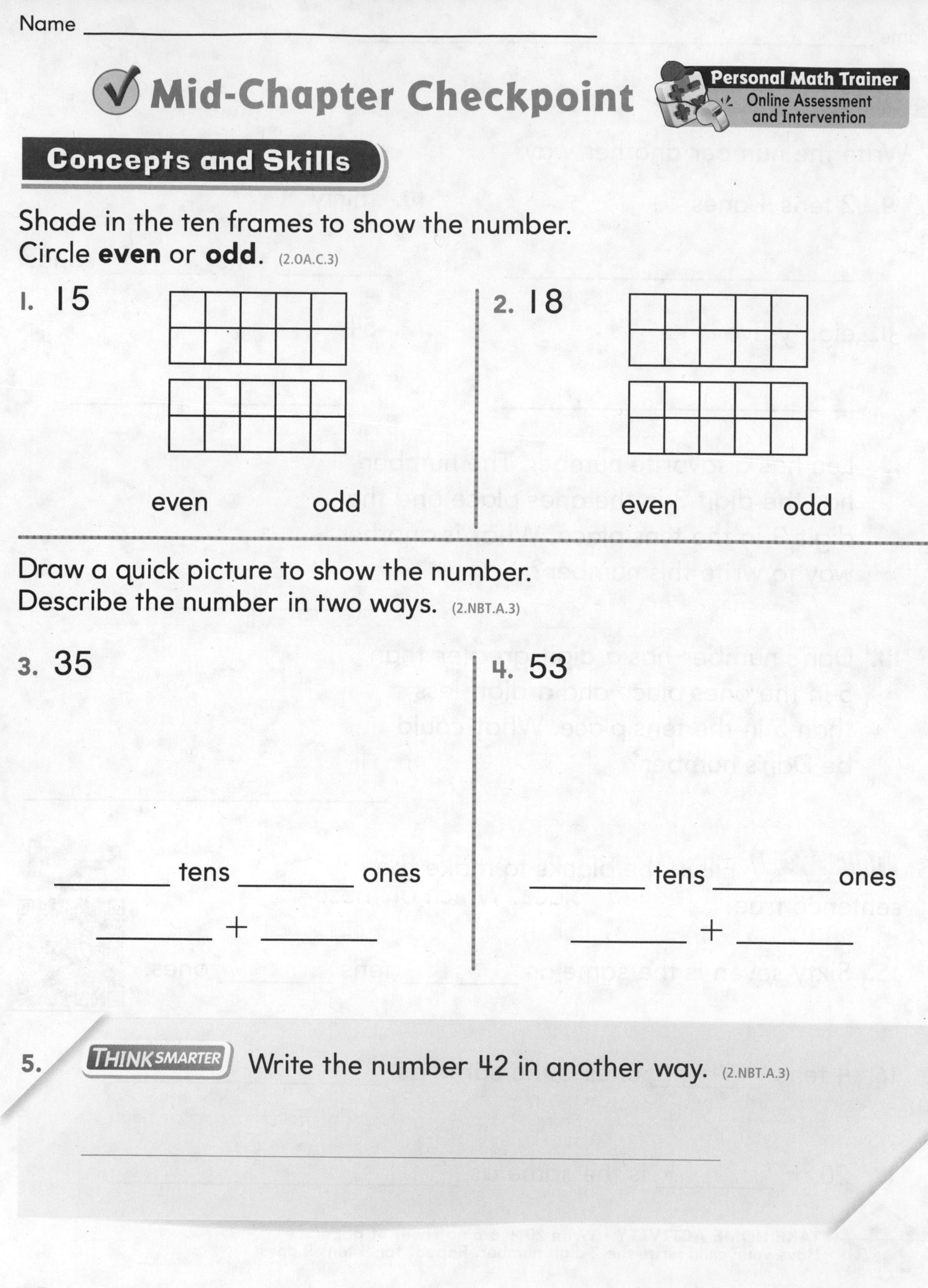

Name ____________________________

Mid-Chapter Checkpoint

Concepts and Skills

Shade in the ten frames to show the number.
Circle **even** or **odd**. (2.OA.C.3)

1. 15

even odd

2. 18

even odd

Draw a quick picture to show the number.
Describe the number in two ways. (2.NBT.A.3)

3. 35

_____ tens _____ ones

_____ + _____

4. 53

_____ tens _____ ones

_____ + _____

5. THINKSMARTER Write the number 42 in another way. (2.NBT.A.3)

____

Name ________________________________

Different Ways to Write Numbers

Common Core **COMMON CORE STANDARD—2.NBT.A.3**
Understand place value.

Write the number another way.

1. 32

_____ tens _____ ones

2. forty-one

____

3. 9 tens 5 ones

____

4. 80 + 3

____

5. 57

_____ tens _____ ones

6. seventy-two

_____ + _____

7. 60 + 4

____

8. 4 tens 8 ones

____

Problem Solving Real World

9. A number has the digit 3 in the ones place and the digit 4 in the tens place. Which of these is another way to write this number? Circle it.

3 + 4   40 + 3   30 + 4

10. WRITE Math Write the number 63 in four different ways.

Lesson Check (2.NBT.A.3)

1. Write 3 tens 9 ones in another way.

____

2. Write the number eighteen in another way.

____

Spiral Review (2.NBT.A.3)

3. Write the number 47 in tens and ones.

_____ tens _____ ones

4. Write the number 95 in words.

____

5. What is the value of the underlined digit? Write the number.

6<u>1</u>

_____

6. What is the value of the underlined digit? Write the number.

<u>1</u>7

_____

FOR MORE PRACTICE GO TO THE Personal Math Trainer

Name ________________________________

Algebra • Different Names for Numbers

Essential Question How can you show the value of a number in different ways?

Common Core Number and Operations in Base Ten—2.NBT.A.3
MATHEMATICAL PRACTICES
MP1, MP6, MP7, MP8

Listen and Draw Real World Hands On

Use [tens] [ones] to show the number different ways.
Record the tens and ones.

________ tens ________ ones

________ tens ________ ones

________ tens ________ ones

FOR THE TEACHER • Read the following problem. Syed has 26 rocks. What are some different ways to show 26 with blocks? Have children start with 26 ones blocks. Then have them use base-ten blocks and record the number of tens and ones in each of their models.

Math Talk MATHEMATICAL PRACTICES 1

Describe how you can use addition to write the number 26.

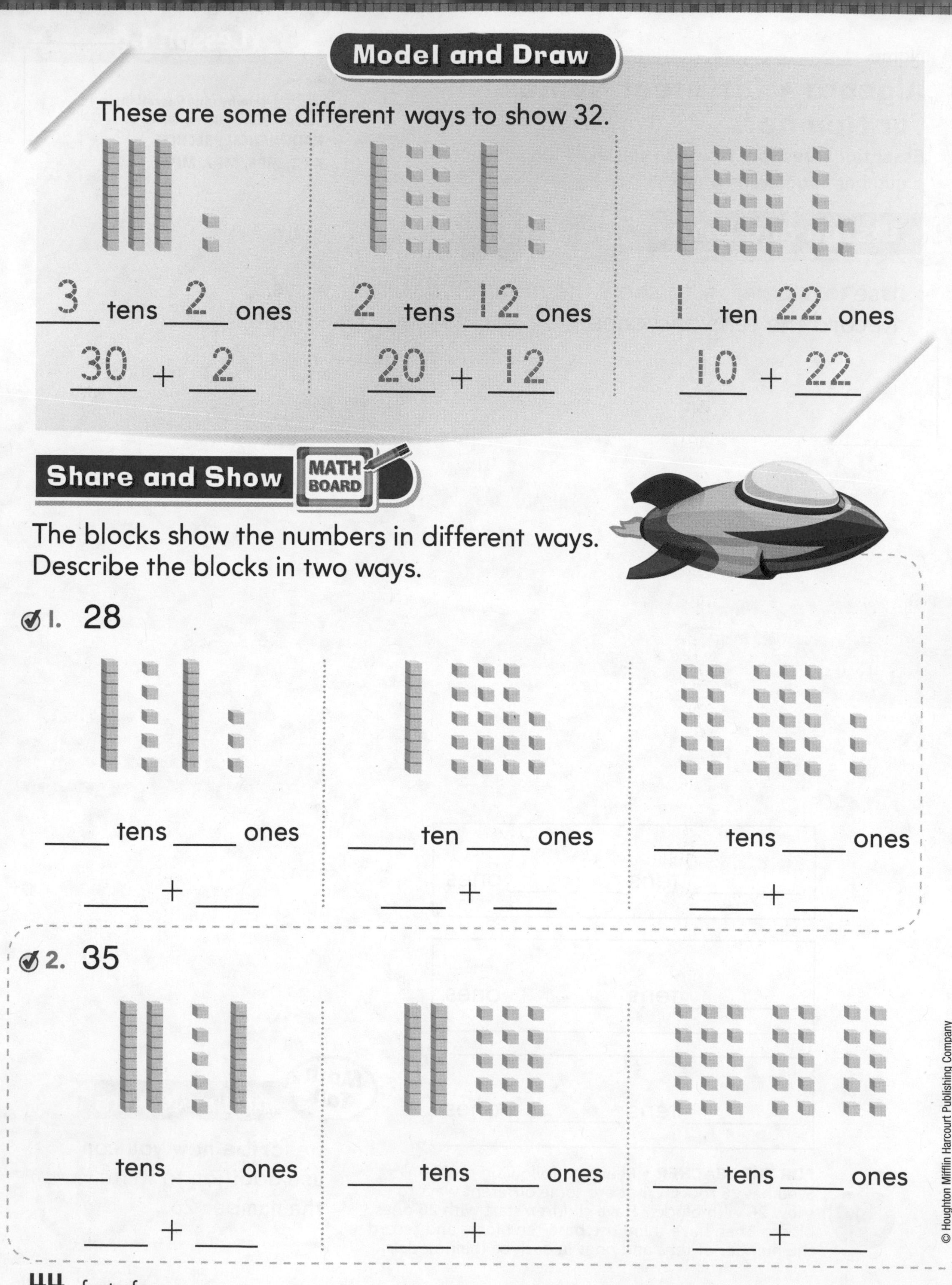

Model and Draw

These are some different ways to show 32.

3 tens 2 ones	2 tens 12 ones	1 ten 22 ones
30 + 2	20 + 12	10 + 22

Share and Show

MATH BOARD

The blocks show the numbers in different ways.
Describe the blocks in two ways.

1. 28

____ tens ____ ones	____ ten ____ ones	____ tens ____ ones
____ + ____	____ + ____	____ + ____

2. 35

____ tens ____ ones	____ tens ____ ones	____ tens ____ ones
____ + ____	____ + ____	____ + ____

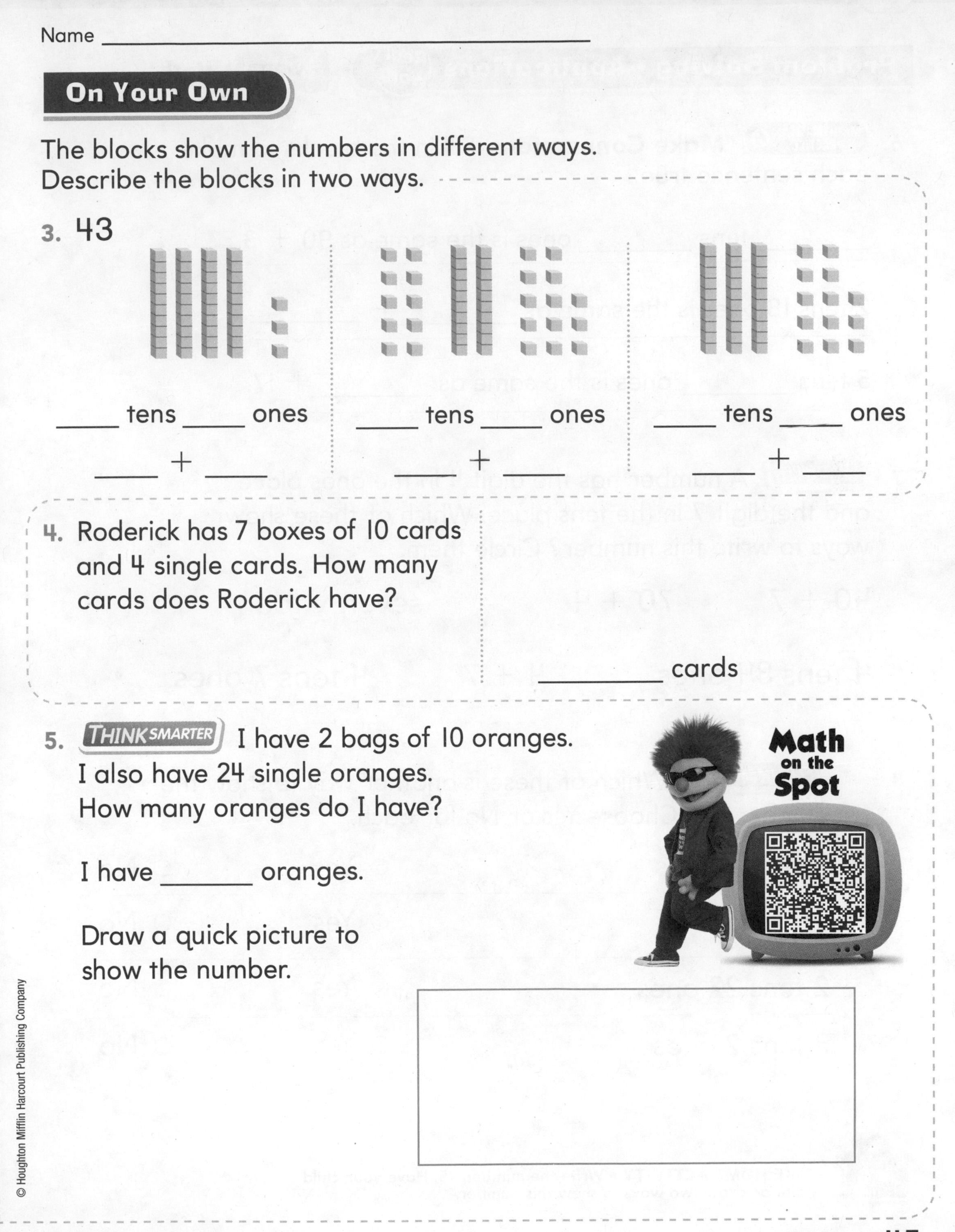

Name ______________________

On Your Own

The blocks show the numbers in different ways.
Describe the blocks in two ways.

3. 43

____ tens ____ ones

____ + ____

____ tens ____ ones

____ + ____

____ tens ____ ones

____ + ____

4. Roderick has 7 boxes of 10 cards and 4 single cards. How many cards does Roderick have?

________ cards

5. THINKSMARTER I have 2 bags of 10 oranges. I also have 24 single oranges. How many oranges do I have?

I have ______ oranges.

Draw a quick picture to show the number.

Problem Solving • Applications Real World

WRITE Math

6. MATHEMATICAL PRACTICE 6 **Make Connections** Fill in the blanks to make each sentence true.

_______ tens _______ ones is the same as $90 + 3$.

2 tens 18 ones is the same as _______ + _______.

5 tens _______ ones is the same as _______ + 17.

7. GO DEEPER A number has the digit 4 in the ones place and the digit 7 in the tens place. Which of these show ways to write this number? Circle them.

$40 + 7$ $70 + 4$ seventy-four

4 tens 34 ones $4 + 7$ 4 tens 7 ones

8. THINK SMARTER Which of these is another way to show the number 42? Choose Yes or No for each.

1 ten 42 ones	○ Yes	○ No
$30 + 12$	○ Yes	○ No
2 tens 22 ones	○ Yes	○ No
3 tens 2 ones	○ Yes	○ No

TAKE HOME ACTIVITY • Write the number 45. Have your child write or draw two ways to show this number.

Name ____________________

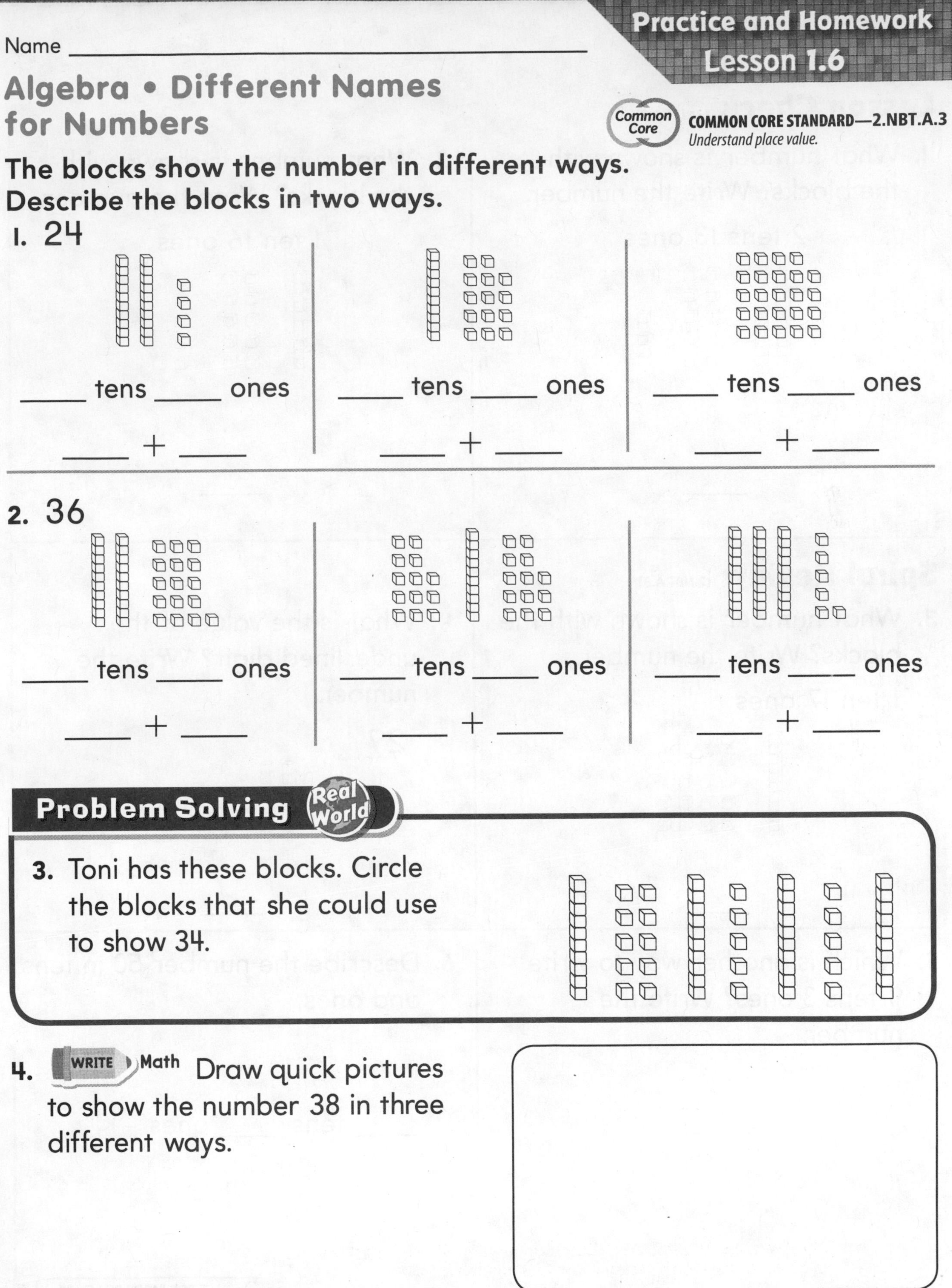

Algebra • Different Names for Numbers

COMMON CORE STANDARD—2.NBT.A.3
Understand place value.

The blocks show the number in different ways. Describe the blocks in two ways.

1. 24

____ tens ____ ones
____ + ____

____ tens ____ ones
____ + ____

____ tens ____ ones
____ + ____

2. 36

____ tens ____ ones
____ + ____

____ tens ____ ones
____ + ____

____ tens ____ ones
____ + ____

Problem Solving — Real World

3. Toni has these blocks. Circle the blocks that she could use to show 34.

4. WRITE Math Draw quick pictures to show the number 38 in three different ways.

Lesson Check (2.NBT.A.3)

1. What number is shown with the blocks? Write the number.

2 tens 13 ones

_____

2. What number is shown with the blocks? Write the number.

1 ten 16 ones

_____

Spiral Review (2.NBT.A.3)

3. What number is shown with the blocks? Write the number.

1 ten 17 ones

_____

4. What is the value of the underlined digit? Write the number.

2<u>9</u>

_____

5. Which is another way to write 9 tens 3 ones? Write the number.

_____

6. Describe the number 50 in tens and ones.

_____ tens _____ ones

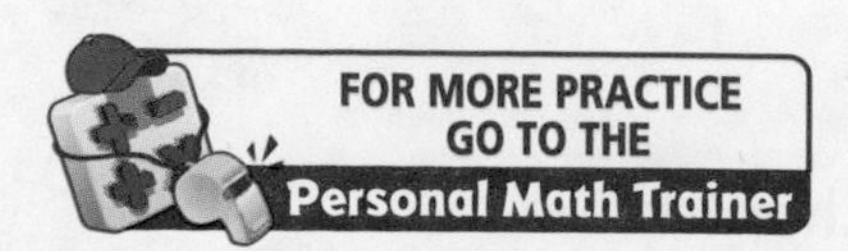

Name ______________________

Problem Solving • Tens and Ones

Essential Question How does finding a pattern help you find all the ways to show a number with tens and ones?

Common Core **Number and Operations in Base Ten—2.NBT.A.3**
MATHEMATICAL PRACTICES MP1, MP4, MP7

Gail needs to buy 32 pencils. She can buy single pencils or boxes of 10 pencils. What are all of the different ways Gail can buy 32 pencils?

Unlock the Problem (Real World)

What do I need to find?

ways Gail can buy 32 pencils

What information do I need to use?

She can buy single pencils or boxes of 10 pencils.

Show how to solve the problem.
Draw quick pictures for 32. Complete the chart.

Boxes of 10 pencils	Single pencils
3	2
2	12
1	
0	

HOME CONNECTION • Your child found a pattern in the different combinations of tens and ones. Using a pattern helps to make an organized list.

Try Another Problem

- What do I need to find?
- What information do I need to use?

Find a pattern to solve.

1. Sara has 36 crayons. She can pack them in boxes of 10 crayons or as single crayons. What are all of the ways Sara can pack the crayons?

Boxes of 10 crayons	Single crayons
3	6

2. Mr. Winter is putting away 48 chairs. He can put away the chairs in stacks of 10 or as single chairs. What are all of the ways Mr. Winter can put away the chairs?

Stacks of 10 chairs	Single chairs
4	8

Math Talk

MATHEMATICAL PRACTICES 7

Look for Structure
Describe a pattern you can use to write the number 32.

Name ______________________________

Share and Show

Find a pattern to solve.

✓3. Philip is putting 25 markers into a bag. He can put the markers in the bag as bundles of 10 or as single markers. What are all of the ways Philip can put the markers in the bag?

Bundles of 10 markers	Single markers

✓4. Stickers are sold in packs of 10 stickers or as single stickers. Miss Allen wants to buy 33 stickers. What are all of the ways she can buy the stickers?

Packs of 10 stickers	Single stickers

5. THINKSMARTER Devin had 32 baseball cards. He gets 7 more cards. He can pack them in boxes of 10 cards or as single cards. What are all of the ways Devin can sort the cards?

Math on the Spot

Boxes of 10 cards	Single cards

On Your Own

Solve. Write or draw to explain.

6. MATHEMATICAL PRACTICE 7 **Look for Structure**
Lee can pack her toy cars in boxes of 10 cars or as single cars. Which of these is a way that she can pack her 24 toy cars? Circle your answer.

4 boxes of 10 cars and 2 single cars	1 box of 10 cars and 24 single cars	2 boxes of 10 cars and 4 single cars

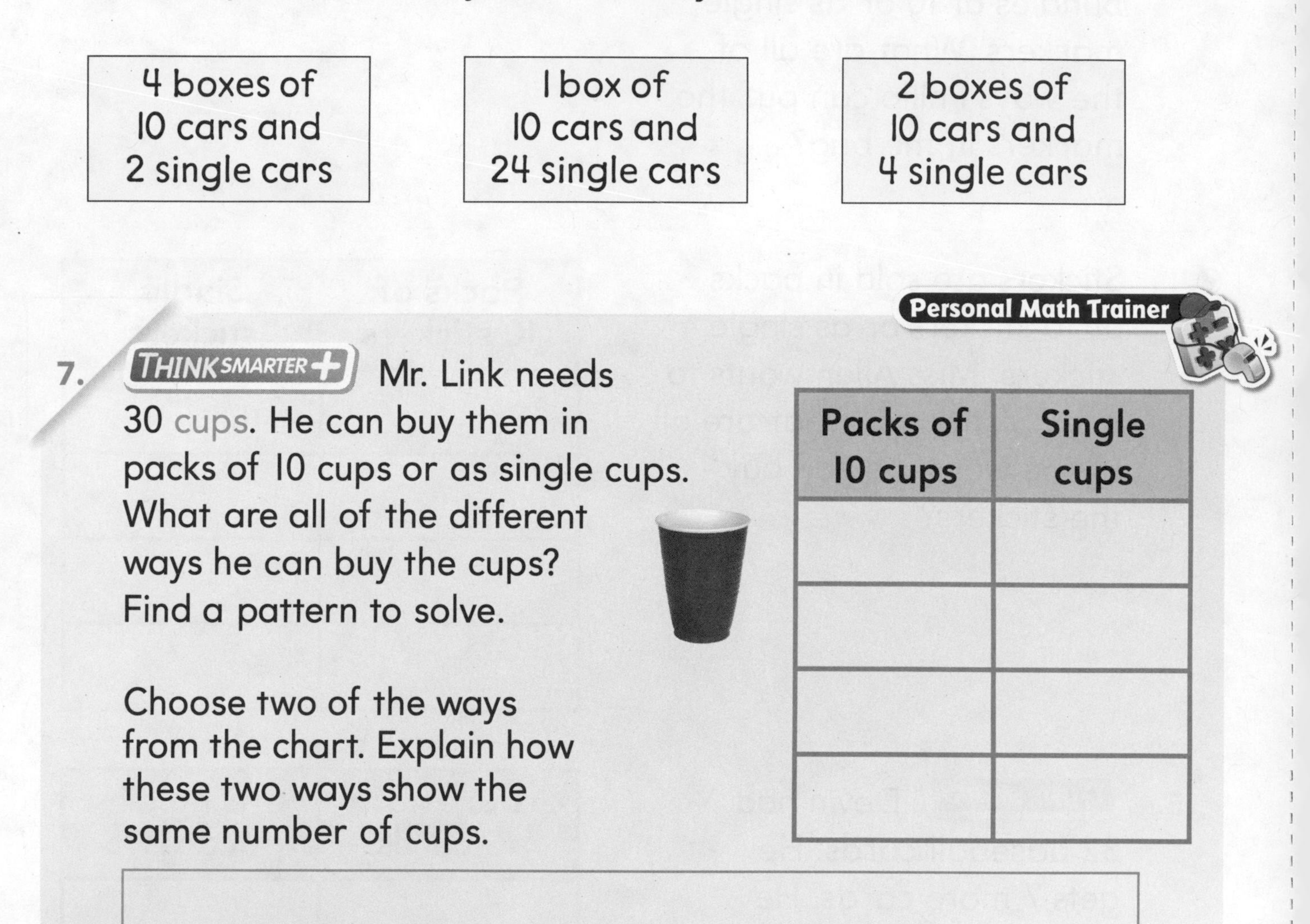

Personal Math Trainer

7. THINKSMARTER+ Mr. Link needs 30 cups. He can buy them in packs of 10 cups or as single cups. What are all of the different ways he can buy the cups? Find a pattern to solve.

Packs of 10 cups	Single cups

Choose two of the ways from the chart. Explain how these two ways show the same number of cups.

TAKE HOME ACTIVITY • Have your child explain how he or she solved one of the exercises in this lesson.

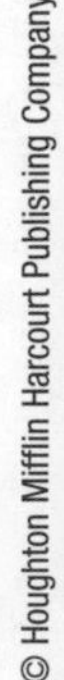

Name ______________________________

Problem Solving • Tens and Ones

COMMON CORE STANDARD—2.NBT.A.3
Understand place value.

Find a pattern to solve.

1. Ann is grouping 38 rocks. She can put them into groups of 10 rocks or as single rocks. What are the different ways Ann can group the rocks?

Groups of 10 rocks	Single rocks

2. Mr. Grant needs 30 pieces of felt. He can buy them in packs of 10 or as single pieces. What are the different ways Mr. Grant can buy the felt?

Packs of 10 pieces	Single pieces

3. WRITE Math Choose one of the problems above. Describe how you organized the answers.

____

____

____

____

____

____

Lesson Check (2.NBT.A.3)

1. Mrs. Chang is packing 38 apples. She can pack them in bags of 10 or as single apples. Complete the table to show another way Mrs. Chang can pack the apples.

Bags of 10 apples	Single apples
2	18
1	28
0	38

Spiral Review (2.NBT.A.3)

2. What is the value of the underlined digit? Write the number.

<u>5</u>4

____

3. What number is shown with the blocks? Write the number.

2 tens 19 ones

____

4. Write the number 62 in words.

____

5. What number can be written as 8 tens and 6 ones? Write the number.

____

FOR MORE PRACTICE GO TO THE **Personal Math Trainer**

Name ____________________

Counting Patterns Within 100

Essential Question How do you count by 1s, 5s, and 10s with numbers less than 100?

Common Core **Number and Operations in Base Ten—2.NBT.A.2**
MATHEMATICAL PRACTICES
MP1, MP3, MP7

Listen and Draw

Look at the hundred chart. Write the missing numbers.

1	2	3		5	6		8		10
11		13	14	15	16		18	19	20
	22	23	24		26	27	28	29	30
31	32		34	35	36		38	39	
41		43	44	45	46	47		49	50
51		53		55		57		59	60
	62		64	65	66	67	68		70
71	72	73	74		76		78	79	
81		83		85	86	87	88	89	90
	92		94	95	96		98		100

Math Talk MATHEMATICAL PRACTICES 1

Describe some different ways to find the missing numbers in the chart.

FOR THE TEACHER • Have children complete the hundred chart to review counting to 100.

Model and Draw

You can count on by different amounts.
You can start counting with different numbers.

Count by ones.

1, 2, 3, 4, 5, 6, ____, ____

29, 30, 31, 32, 33, ____, ____, ____

Count by fives.

5, 10, 15, 20, ____, ____, ____, ____

50, 55, 60, 65, ____, ____, ____, ____

Share and Show

MATH BOARD

Count by ones.

1. 15, 16, 17, ____, ____, ____, ____, ____

Count by fives.

2. 15, 20, 25, ____, ____, ____, ____, ____

✓3. 60, 65, ____, ____, ____, ____, ____

Count by tens.

4. 10, 20, ____, ____, ____, ____, ____

✓5. 30, 40, ____, ____, ____, ____, ____

Name ______________________

On Your Own

Count by ones.

6. 77, 78, ______, ______, ______, ______, ______

Count by fives.

7. 35, 40, ______, ______, ______, ______, ______

Count by tens.

8. 20, 30, ______, ______, ______, ______, ______

9. Amber counts by fives to 50.
How many numbers will she say?

______ numbers

10. THINKSMARTER Dinesh counts by fives to 100.
Gwen counts by tens to 100.
Who will say more numbers? Explain.

____

____

____

____

____

Math on the Spot

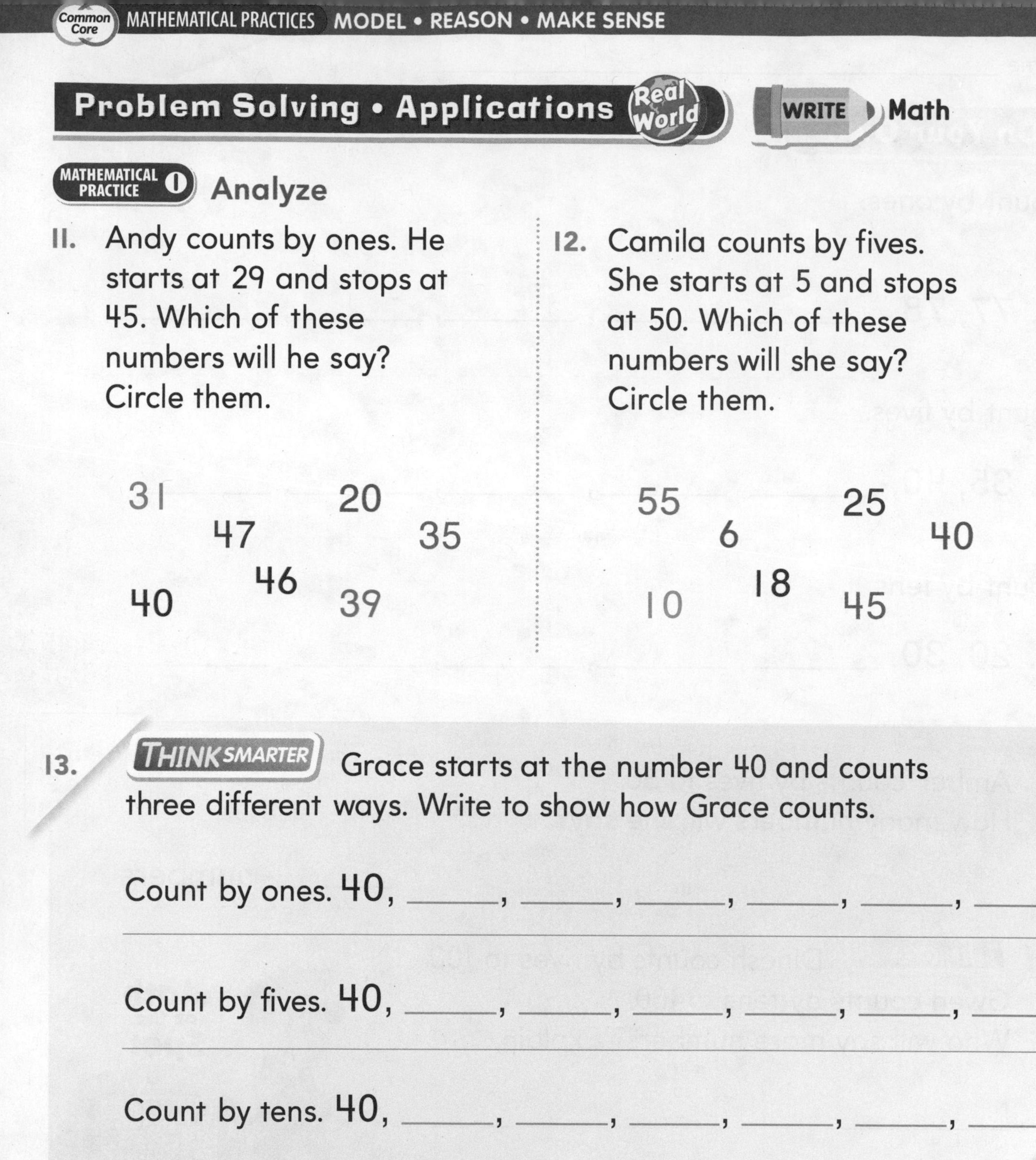

Problem Solving • Applications Real World

WRITE Math

MATHEMATICAL PRACTICE 1 **Analyze**

11. Andy counts by ones. He starts at 29 and stops at 45. Which of these numbers will he say? Circle them.

31 20 47 35 46 40 39

12. Camila counts by fives. She starts at 5 and stops at 50. Which of these numbers will she say? Circle them.

55 25 6 40 18 10 45

13. THINK SMARTER Grace starts at the number 40 and counts three different ways. Write to show how Grace counts.

Count by ones. 40, ______, ______, ______, ______, ______, ______

Count by fives. 40, ______, ______, ______, ______, ______, ______

Count by tens. 40, ______, ______, ______, ______, ______, ______

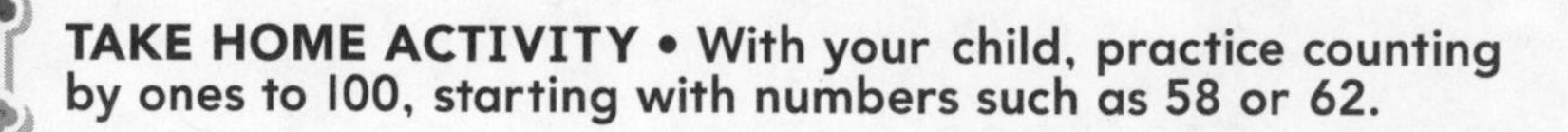

TAKE HOME ACTIVITY • With your child, practice counting by ones to 100, starting with numbers such as 58 or 62.

Name ________________________________

Counting Patterns Within 100

Common Core **COMMON CORE STANDARD—2.NBT.A.2**
Understand place value.

Count by ones.

1. 58, 59, ____, ____, ____, ____, ____

Count by fives.

2. 45, 50, ____, ____, ____, ____, ____

3. 20, 25, ____, ____, ____, ____, ____

Count by tens.

4. 20, ____, ____, ____, ____, ____, ____

Count back by ones.

5. 87, 86, 85, ____, ____, ____

Problem Solving Real World

6. Tim counts his friends' fingers by fives. He counts six hands. What numbers does he say?

5, ____, ____, ____, ____, ____

7. WRITE Math Count by 1s or 5s. Write the first five numbers you would count, starting at 15.

Lesson Check (2.NBT.A.2)

1. Count by fives.

70, ____, ____, ____, ____

2. Count by tens.

60, ____, ____, ____, ____

Spiral Review (2.OA.C.3, 2.NBT.A.2, 2.NBT.A.3)

3. Count back by ones.

21, ____, ____, ____, ____

4. A number has 2 tens and 15 ones. Write the number in words.

____

5. Describe the number 72 in tens and ones.

____ tens ____ ones

6. Find the sum. Is the sum even or odd? Write even or odd.

$9 + 9 =$ ____

____

FOR MORE PRACTICE GO TO THE **Personal Math Trainer**

Name ______________________________

Counting Patterns Within 1,000

Essential Question How do you count by 1s, 5s, 10s, and 100s with numbers less than 1,000?

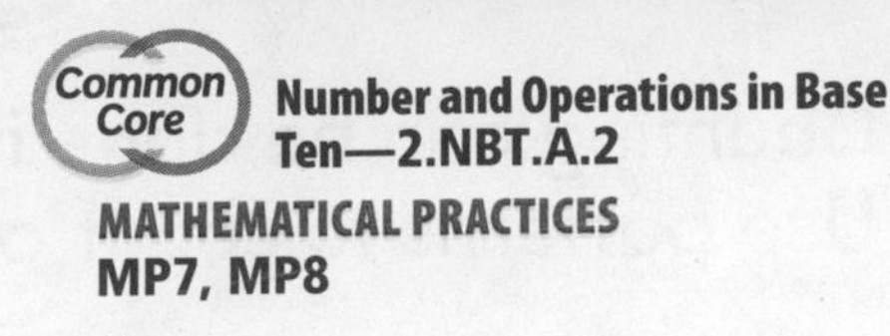

Number and Operations in Base Ten—2.NBT.A.2

MATHEMATICAL PRACTICES
MP7, MP8

Listen and Draw

Write the missing numbers in the chart.

401		403	404		406	407	408		410
411				415	416	417	418	419	
421	422	423	424	425		427	428	429	430
	432		434	435	436	437	438		
441	442	443	444		446	447		449	450
			454	455	456	457	458	459	460
461	462						468	469	470
	472	473	474	475	476	477		479	480
481	482		484	485	486				490
	492	493		495	496	497	498		

FOR THE TEACHER • Have children complete the number chart to practice counting with 3-digit numbers.

Math Talk MATHEMATICAL PRACTICES 7

Look for Structure
What counting patterns could you use to complete the chart?

Model and Draw

Counting can be done in different ways.
Use patterns to count on.

Count by fives.

95, 100, 105, 110, 115, ______, ______

140, 145, 150, 155, ______, ______, ______

Count by tens.

300, 310, 320, ______, ______, ______, ______

470, 480, 490, ______, ______, ______, ______

Share and Show MATH BOARD

Count by fives.

1. 745, 750, 755, ______, ______, ______, ______

Count by tens.

2. 520, 530, 540, ______, ______, ______, ______

3. 600, 610, ______, ______, ______, ______, ______

Count by hundreds.

4. 100, 200, ______, ______, ______, ______, ______

5. 300, 400, ______, ______, ______, ______, ______

Name ______________________

On Your Own

Count by fives.

6. 215, 220, 225, ______, ______, ______, ______

7. 905, 910, ______, ______, ______, ______, ______

Count by tens.

8. 730, 740, 750, ______, ______, ______, ______

9. 160, 170, ______, ______, ______, ______, ______

Count by hundreds.

10. 200, 300, ______, ______, ______, ______, ______

11. THINK SMARTER Martin starts at 300 and counts by fives to 420. What are the last 6 numbers Martin will say?

______, ______, ______, ______, ______, ______

12. The book fair has 390 books. They have 5 more boxes with 10 books in each box. Count by tens. How many books are at the book fair.

______ books

Math on the Spot

Problem Solving • Applications Real World

WRITE Math

MATHEMATICAL PRACTICE 7 **Look for a Pattern**

13. Lisa counts by fives. She starts at 120 and stops at 175. Which of these numbers will she say? Circle them.

170 151 135

155 200 180

14. George counts by tens. He starts at 750 and stops at 830. Which of these numbers will he say? Circle them.

755 690 780

760 795 810

15. THINK SMARTER Carl counts by hundreds. Which of these show ways that Carl could count? Choose Yes or No for each.

100, 110, 120, 130, 140	○ Yes	○ No
100, 200, 300, 400, 500	○ Yes	○ No
500, 600, 700, 800, 900	○ Yes	○ No
300, 305, 310, 315, 320	○ Yes	○ No

TAKE HOME ACTIVITY • With your child, count by fives from 150 to 200.

Name ______________________________

Counting Patterns Within 1,000

Common Core **COMMON CORE STANDARD—2.NBT.A.2**
Understand place value.

Count by fives.

1. 415, 420, ______, ______, ______, ______

2. 675, 680, ______, ______, ______, ______, ______

Count by tens.

3. 210, 220, ______, ______, ______, ______, ______

Count by hundreds.

4. 300, 400, ______, ______, ______, ______, ______

Count back by ones.

5. 953, 952, ______, ______, ______, ______, ______

Problem Solving Real World

6. Lee has a jar of 100 pennies.
   She adds groups of 10 pennies to the jar.
   She adds 5 groups. What numbers does she say?

   ______, ______, ______, ______, ______

7. WRITE Math Count by fives from 135 to 175. Write these numbers and describe the pattern.

____

____

Lesson Check (2.NBT.A.2)

1. Count by tens.

160, ____, ____, ____, ____

2. Count by hundreds.

400, ____, ____, ____, ____

Spiral Review (2.NBT.A.2, 2.NBT.A.3)

3. Count by fives.

245, ____, ____, ____, ____

4. Count back by ones.

71, ____, ____, ____, ____

5. Describe 45 in another way.

____ tens ____ ones

6. Describe 7 tens 9 ones in another way.

____

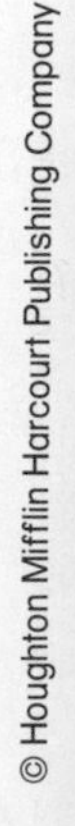

FOR MORE PRACTICE GO TO THE **Personal Math Trainer**

Name ______________________

Chapter 1 Review/Test

Personal Math Trainer
Online Assessment and Intervention

1. Does the ten frame show an even number? Choose Yes or No.

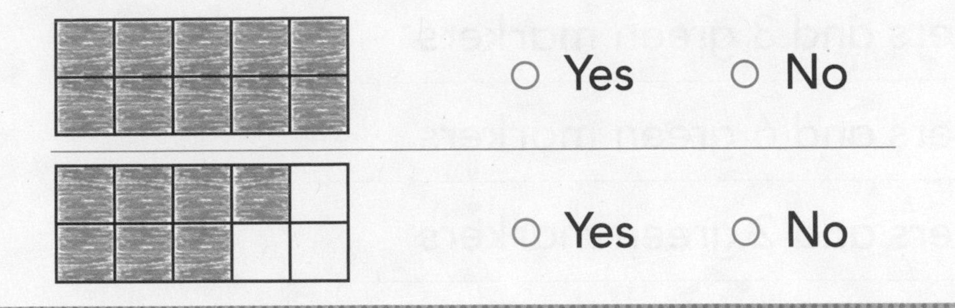

○ Yes ○ No

○ Yes ○ No

2. Write an even number between 7 and 16. Draw a picture and then write a sentence to explain why it is an even number.

____

____

3. What is the value of the digit 5 in the number 75?

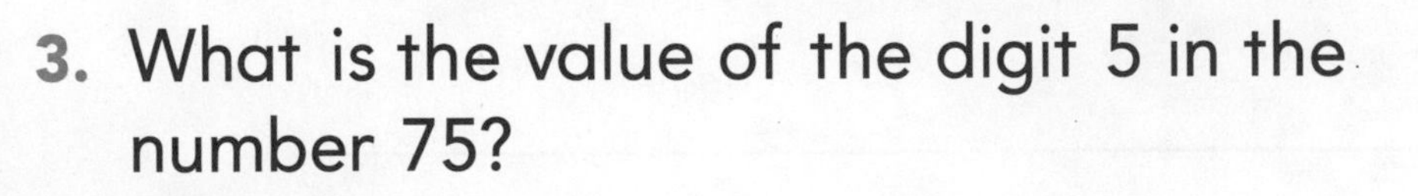

______

4. GO DEEPER Ted has an even number of yellow markers and an odd number of green markers. Choose all the groups of markers that could belong to Ted.

- ○ 8 yellow markers and 3 green markers
- ○ 3 yellow markers and 6 green markers
- ○ 4 yellow markers and 2 green markers
- ○ 6 yellow markers and 7 green markers

5. Jeff starts at 190 and counts by tens. What are the next 6 numbers Jeff will say?

190, ______, ______, ______, ______, ______, ______

6. Megan counts by ones to 10. Lee counts by fives to 20. Who will say more numbers? Explain.

____

____

____

Name ______________________

7. Draw a picture to show the number 43.

Describe the number 43 in two ways.

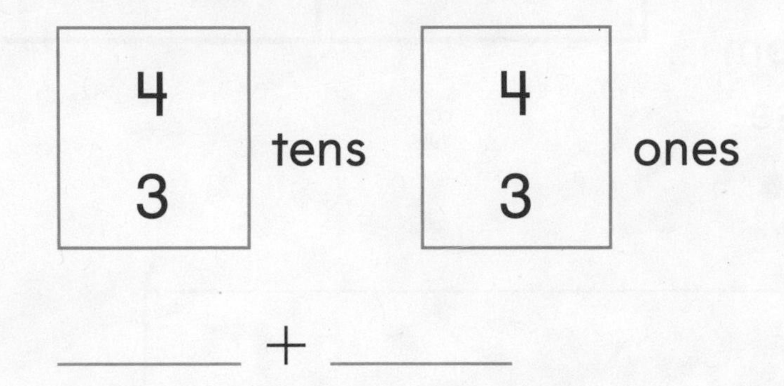

_____ + _____

8. Jo lives on Maple Road. Her address has the digit 2 in the ones place and the digit 4 in the tens place. What is Jo's address?

_____ Maple Road

9. Do the numbers show counting by fives? Choose Yes or No.

76, 77, 78, 79, 80	○ Yes	○ No
20, 30, 40, 50, 60	○ Yes	○ No
70, 75, 80, 85, 90	○ Yes	○ No
35, 40, 45, 50, 55	○ Yes	○ No

Personal Math Trainer

10. THINKSMARTER+ Mrs. Payne needs 35 notepads. She can buy them in packs of 10 notepads or as single pads. What are all the different ways Mrs. Payne can buy the notepads? Find a pattern to solve.

Packs of 10 notepads	Single notepads

Choose two of the ways from the chart. Explain how these two ways show the same number of notepads.

11. Ann has a favorite number. It has a digit less than 4 in the tens place. It has a digit greater than 6 in the ones place. Could the number be Ann's number? Choose Yes or No.

$30 + 9$	○ Yes	○ No
sixty-seven	○ Yes	○ No
2 tens 8 ones	○ Yes	○ No

Write another number that could be Ann's favorite. ______